The Grammar of Pattern

Textile Institute Professional Publications

Series Editor: The Textile Institute

Care and Maintenance of Textile Products Including Apparel and Protective Clothing
Rajkishore Nayak, Saminathan Ratnapandian

The Grammar of Pattern
Michael Hann

For more information about this series, please visit: www.crcpress.com/Textile-Institute-Professional-Publications/book-series/TIPP

The Grammar of Pattern

Michael Hann

CRC Press
Taylor & Francis Group
Boca Raton London New York

CRC Press is an imprint of the
Taylor & Francis Group, an **informa** business

The Textile Institute

CRC Press
Taylor & Francis Group
6000 Broken Sound Parkway NW, Suite 300
Boca Raton, FL 33487-2742

© 2019 by Taylor & Francis Group, LLC
CRC Press is an imprint of Taylor & Francis Group, an Informa business

No claim to original U.S. Government works

Printed on acid-free paper

International Standard Book Number-13: 978-1-138-06558-1 (Paperback)
978-1-138-06561-1 (Hardback)

**Visit the Taylor & Francis Web site at
http://www.taylorandfrancis.com**

**and the CRC Press Web site at
http://www.crcpress.com**

CONTENTS

Contents

LIST OF ILLUSTRATIONS

Note: It should be noted that major contributions to the production of illustrations were made by Chaoran Wang (referred to as CW below), Haohong Zhuang (referred to as HZ) and Jia Zhuang (referred to as JZ).

SERIES PREFACE

The aim of the *Textile Institute Professional Publications* is to provide support to textile professionals in their work and to help emerging professionals, such as final year or Master's students, by providing the information needed to gain a sound understanding of key and emerging topics relating to textile, clothing and footwear technology, textile chemistry, materials science, and engineering. The books are written by experienced authors with expertise in the topic and all texts are independently reviewed by textile professionals or textile academics.

The textile industry has a history of being both an innovator and an early adopter of a wide variety of technologies. There are textile businesses of some kind operating in all counties across the world. At any one time, there is an enormous breadth of sophistication in how such companies might function. In some places where the industry serves only its own local market, design, development, and production may continue to be based on traditional techniques, but companies that aspire to operate globally find themselves in an intensely competitive environment, some driven by the need to appeal to followers of fast-moving fashion, others by demands for high performance and unprecedented levels of reliability. Textile professionals working within such organisations are subjected to a continued pressing need to introduce new materials and technologies, not only to improve production efficiency and reduce costs, but also to enhance the attractiveness and performance of their existing products and to bring new products into being. As a consequence, textile academics and professionals find themselves having to continuously improve their understanding of a wide range of new materials and emerging technologies to keep pace with competitors.

The Textile Institute was formed in 1910 to provide professional support to textile practitioners and academics undertaking research and teaching in the field of textiles. The Institute quickly established itself as the professional body for textiles worldwide and now has individual and corporate members in over 80 countries. The Institute works to provide sources of reliable and up-to-date information to support textile professionals through its research journals, the *Journal of the Textile Institute*[1] and *Textile Progress*[2], definitive descriptions of textiles and their components through its online publication *Textile Terms and Definitions*[3], and contextual treatments of important topics within the field of textiles in the form of self-contained books such as the *Textile Institute Professional Publications*.

REFERENCES

1. http://www.tandfonline.com/action/journalInformation?show=aimsScope&journalCode=tjti20
2. http://www.tandfonline.com/action/journalInformation?show=aimsScope&journalCode=ttpr20
3. http://www.ttandd.org

PREFACE

The title adopted for this present book is a tribute to the endeavours of Owen Jones and his monumental treatise *The Grammar of Ornament,* published in 1856, five years after the Great Exhibition, an event that highlighted the dearth of effective design in Victorian Britain. Jones's publication included a series of impressive chromolithographic plates showing examples of visual arts, including regular patterns from several cultures, historical and contemporary, as well as a series of propositions claiming to codify various universal principles of good design. *The Grammar of Ornament* was a great stimulus to British manufacturers and, after its publication in French and German, similar compendia illustrating regular repeating patterns and other visual arts from a multitude of sources were published across Europe. Reviews of Jones's contribution and the importance of the Great Exhibition of 1851 in terms of its influence on design were provided by Jespersen (2008) and Sloboda (2008).

While the title of this present book makes close reference to the title of Jones's renowned treatise, it is recognised that the use of the word 'grammar' suggests belief in a relationship between the rules governing a language and those governing regular-pattern composition and design. There are however past researchers who have not approved of drawing such parallels. Dondis (1973, p. 9), for example, commented on the futility of attempting to draw parallels between language and visual literacy, maintaining that languages were 'logical wholes', and that no such characteristic could 'be ascribed to visual intelligence'. So, Dondis believed that drawing an analogy between the rules governing a language and the visual arts was a wasted exercise. A contrary view was, however, taken by Westphal-Fitch et al. (2012), who claimed that 'visual patterns' had characteristics 'in common with music and language' and were 'governed by a set of combinatorial principles' or 'grammars' that constrained 'the arrangement of units into groups on multiple hierarchical levels'. They commented further that although 'formal language theory is most typically used in the context of linear sequences or strings (e.g. in linguistics, computer programming and molecular biology), it can be naturally extended to cover two-dimensional patterns as well'. Similarly, Leborg (2004, p. 5), in his treatise entitled *Visual Grammar,* commented that the reason for 'writing a grammar of visual language is the same for any language: to define its basic elements, describe its patterns and processes, and to understand the relations between the individual elements in the system'. Knight (1993), too, considered the use of the word 'grammar' when referring to the visual arts and observed that 'scholars and practitioners in architecture and the arts have used the concepts of 'grammar' and language to characterise in an intuitive way the principles or rules underlying designs in different styles' and observed further that 'a powerful and compelling formation for describing architectural and other designs, called a shape grammar' had been developed. The present author agrees with Westphal-Fitch et al. (2012), Leborg (2004) and Knight (1993): the comparison with grammar is an effective means of breaking visual entities into their basic structural component parts and is a useful procedure to gain an intimate understanding of all visual statements. However, it is not the intention in this present book to explore this area further. Rather, it is stressed simply that the use of the word 'grammar' in the title of this present book is, on the one hand, an acknowledgement of the vast contribution made by Jones (1986 [1856]), and, on the other hand, an expression of the desire to identify some of the components and rules governing regular patterns in terms of their composition and construction.

The publication of *The Grammar of Ornament* was instrumental in opening the debate concerning design practice and education in Britain. Many educational developments and the initiation of numerous debates relating to the role of the visual arts and design and their part in wider society can be traced to that time and later. Brett (1988), for example, provided an outline of the nineteenth-century debate between Ruskin and Dresser, both eminent theoreticians of the day, over the preferred path of drawing instruction and the development of drawing skills in British design schools. These matters are not, however, the immediate concern of the present book. Rather, the aim is to develop an understanding among readers of the characteristics of regular patterns. Complementing this, guidelines for original composition and design are proposed using various frameworks (referred to as 'grids') as templates to the placing of parts. The book identifies, illustrates and explains the structure of a wide range of pattern types, including patterns which feature regular distributions of spots, stripes, medallions and checks, as well as regular all-over patterns known as 'diapers'. Systems of regular-pattern classification, announced initially by Victorian designers such as Lewis Foreman Day, and 'rediscovered' by various crystallographers and mathematicians during the twentieth century, are explained and illustrated. Frameworks for examining symmetry content are reviewed, and various grid forms derived from well-known tiling classes, as well as brick bonds, stained-glass arrangements and Chinese lattice designs, are introduced as guides to assist with pattern construction. Although the focus is on two-dimensional phenomena, the contents should prove of interest also to practitioners and students across the full range of visual-arts-and-design disciplines.

In the early twenty-first century, interest in regular-pattern composition and design developed beyond its traditional home among textile and wallpaper designers to be of interest also to graphic and visual-communication designers, as well as fashion, product, interior and architectural designers. There has thus been the recognition that knowledge of patterns and their construction is of importance to three-dimensional, as well as two-dimensional, design. This book develops an understanding among readers of the characteristics of regular patterns. The intended audience is principally undergraduate visual-arts-and-design students, though it is believed that the content will be of value also to teachers and practitioners, as well as researchers.

Complementing the explanatory nature of this present book, around 200 original black-and-white images (many produced in-house at the University of Leeds), together with thirty-nine previously unpublished colour illustrations (of items held at ULITA – an Archive of International Textiles – at the University of Leeds, UK) are included. Attention is focused on the vast diversity of pattern types possible, often using a limited number of structural rules.

Grids used in the visual arts and design, including regular-pattern design, may be referred to as 'compositional grids'; in the context of regular patterns, these consist of repeating unit cells, with each cell acting as housing for a repeating unit or a component of a repeating unit. A compositional grid acts, therefore, as a guideline for the placement of a regular pattern's component parts; such grids may be apparent or hidden in the regular pattern's final depiction.

In the early twenty-first century, there was a tendency among visual-arts-and-design students to believe that, with the onset of sophisticated digital technology and associated computer-enhanced design possibilities, there was no longer the need to become familiar with visual fundamentals. A contrary view is taken here, and it is stressed that without a well-developed knowledge of visual fundamentals, the basic ingredients of any visual composition, the productive use and exploitation of early twenty-first century computer graphics tools is not assured. In other words, it is emphasised that without a basic understanding of the fundamentals of visual composition, the fruitful exploitation of sophisticated digital-graphics software does not follow. The intention with all the exercises presented at the end of each chapter is to produce images on paper (or other physical surfaces). Often the intended size of regular patterns or other images is not communicated to the viewer. So, a digital image on its own is insufficient to meet the needs of each exercise. It is assumed, therefore, that the intended size is that printed on paper, unless it is suggested otherwise (either by stating an intended scale or by 'mapping' an image onto an intended product using appropriate software). The regular patterns presented in this present book are intended for illustrative purposes only, so anticipated end use, intended scale and method of production are not given.

REFERENCES

Brett, D. 1988. The interpretation of ornament. *Journal of Design History*, 1 (2): 103–111.

Dondis, D. A. 1973. *A Primer of Visual Literacy*. Cambridge (Mass.): MIT Press.

Jespersen, J. K. 2008. Originality and Jones' The Grammar of Ornament of 1856. *Journal of Design History*, 21 (2): 143–153.

Jones, O. 1986 [1856]. *The Grammar of Ornament*. London: Omega and previously London: Day and Son.

Knight, T. W. 1993. Color grammars: The representation of form and color. *Leonardo*, 26 (2): 117–124.

Leborg, C. 2004. *Visual Grammar*. New York: Princeton Architectural Press.

Sloboda, S. 2008. The Grammar of Ornament: Cosmopolitanism and reform in British design. *Journal of Design History*, 21 (3): 223–236.

Westphal-Fitch, G., L. Huber, J. C. Gómez and W. Tecumseh Fitch. 2012. Production and perception rules underlying visual patterns: Effects of symmetry and hierarchy. *Philosophical Transactions: Biological Sciences*, 367 (1598): 2007–2022.

ACKNOWLEDGEMENTS

The author is indebted to I. S. Moxon for his constructive review, useful advice and helpful commentary. Significant contributions to the development of illustrative material were made by Chaoran Wang (CW), Haohong Zhuang (HZ) and Jia Zhuang (JZ). The skills of Keith Findlater ensured that illustrative material was to the standards of quality and definition demanded by the publishers. Thanks are due also to R. Murray and J. Power, both affiliated to the Textile Institute, and to A. Shatkin and colleagues at Taylor and Francis. Acknowledgement is extended to staff and students at Shanghai International College of Fashion Innovation at Donghua University (Shanghai) and to colleagues at the University of Leeds for encouragement and good wishes. Thanks are due also to Jill Winder and Kirsten Doble, both of ULITA – an Archive of International Textiles at the University of Leeds. Acknowledgement and thanks are extended to Peter Dingley for permission to publish images of the work of the designer Guido Marchini and for the provision of enlightening biographical material of the designer while working in the United Kingdom. Various student contributors include Cindy Chao, Hsieh Hsin Ying, Li Yu Ling, Nancy Dai Yi Jie and Rachel H. Leung (of Asia University, Taichung, Taiwan). Thanks are extended also to Zhong Hong and Xun Lin and to the contributions made by past students at the University of Leeds to my improved understanding and appreciation of regular patterns. For past inspiration, acknowledgement is extended to K. P. Hann, M. and J. Large, M. and B. O'Neill, E. and T. Hann, R. and T. Mason, G. M. Thomson, K.C. Jackson. P. T. Speakman, I. Holme, P. Grossberg, J. W. Bell, M. Dobb, J. A. Smith, C. Whewell, D. Young, B. Pourdeyimi, P. Turnbull, C. Hammond, R. McTurk, P. MacGowan, D. O'Neill, E. Broug, D. Holdcroft, A. Watson, H. Coleman, B. Whitaker, J. Rosenthal, H. Hubbard, H. Mee, D. Washburn, D. Crowe, D. Huylebrouck, R. Matteus-Berr, C. Bier, M. Rossi, J. Boomer, T. Ross, A. Hollas and P. Byrne. Last and by no means least, thanks are extended to Naeema B. Hann, Ellen-Ayesha Hann and Haleema-Clare Hann for their unstinting support during the process of manuscript preparation. While every effort has been made to extend thanks where it is required, the author gives unreserved apologies in advance should such acknowledgement be absent.

Michael Hann, Leeds
October 2018

AUTHOR

Professor Michael Hann (BA, MPhil, PhD, FRSA, FRAS, FTI) holds the Chair of Design Theory at the University of Leeds. He is also Director of ULITA – an Archive of International Textiles, an important international archive (and, in the context of this book, a source of illustrative material). He has published across a wide range of subject areas, has made numerous keynote addresses at international conferences, and is an acknowledged international authority on the geometry of design. Recent book publications include: Hann, M (2012). *Structure and Form in Design* (London: Berg); Hann, M. (2013), *Symbol, Pattern and Symmetry* (London: Bloomsbury) and Hann, M. (2015), *Stripes, Grids and Checks* (London: Bloomsbury). He has held adjunct, visiting or invited professorships at institutions in Belgium, Taiwan, Hong Kong, Korea and the Peoples' Republic of China.

CHAPTER 1

PREDOMINANCE

1.1 INTRODUCTION

The passing of the seasons, the phases of the moon and the tides of the sea all adhere to patterns. Patterns permit the birth, life, growth and death of every living creature and the existence of all other natural phenomena. A pattern or series of procedures is required in the manufacture of all products, from pins to hammers, motorcycles to airplanes, books to pens and computers to cell phones. The vast bulk of textiles is produced by manufacturing techniques, each operating to a given series or pattern of actions. But the principal intention here is not to explore these. Rather, the focus is on products produced and how it is often the case that many of these have been designed to include a visual embellishment in the form of a regularly repeating pattern (or 'regular pattern', the term that will be employed subsequently), invariably aimed at an end use and manufactured using a particular raw-material type, colour palette and arrangement of motifs which may hold a particular significance or symbolism to those who produce or use the textile. So, the focus of this book is on regular patterns on textiles and other two-dimensional surfaces and their explanation, as well as on the presentation of compositional formats or arrangements developed over past centuries, though, in many cases, not intended specifically for use in a textile or related context. The content should be of interest not just to textile-design students, practitioners and theorists, but also to those with specialist design interests outside textiles.

It must be stressed at this stage that a regular pattern is a design with an effect which is not characteristic of a non-repeating form of embellishment. Rather, it takes one of two forms: either as a regular border pattern, which extends between two (often) imaginary parallel lines or, alternatively, as a regular all-over pattern which extends across the plane with repetition in two independent directions. A similar differentiation was made by Christie (1969 [1910], p. 1), who was concerned with the evolution of regular patterns and viewed these as designs 'composed of one or more devices, multiplied and arranged in orderly sequence'. At the same time, he accepted that a 'single device, however complicated or complete in itself it may be, is not a pattern, but a unit with which the designer, working according to some definite plan of action, may compose a pattern'. Repetition is an important characteristic of both regular border patterns and regular all-over patterns.

By the early twenty-first century, repetition was a dominant environmental feature, and human beings lived an existence dominated by patterns – visual, audio, mechanical, electronic and environmental. As noted above, regular visual patterns are the concern here, and these may be of two broad classes: regular border patterns, where repetition of a unit is in one direction across the plane, between two imaginary or real parallel lines and regular all-over patterns, where repetition of a unit is in two independent directions and thus covers the two-dimensional plane. With both pattern categories, due to their regular repeating nature, extension to infinity can be imagined. The emphasis in this book will be on regular all-over patterns, though occasional mention and explanation will be given also of regular border patterns, as well as motifs of various kinds (often the building blocks, either on their own or in combination, of the two pattern classes mentioned above).

Terminology presents difficulties, especially when the same term is used by different authors (and, occasionally, by the same author) to refer to different phenomena. In the case of this present book, the more common application is adopted and, where necessary, relevant terms are defined. There are however two exceptions to the use of terminology which need to be explained. When examining the nature of patterns and reviewing relevant literature, it is found often that the words 'ornament' and 'decoration' are used, particularly in twentieth-century literature. However, these terms, 'ornament' and 'decoration', had acquired a derogatory undertone based on the belief that surface additions of any kind were not required for 'good' design and, thus, fell out of favour during the latter part of the twentieth century. During much of the twentieth century, due partly to the influence of the Bauhaus and twentieth-century modernist perspectives on the visual arts and design, embellishment of design was not approved and was perceived as an unnecessary addition (Trilling 2003). Bier (2008) presented a well-focused outline of the debate and, by way of example, highlighted the 'pervasiveness' of pattern in Islamic visual arts 'in virtually all media, at all times', and maintained that such embellishments had more than 'just decorative intent'. Similar perspectives were taken by Bier (1994) in her review of Grabar's (1992) treatise *The Mediation of Ornament*, where she suggested that a more appropriate title of one chapter was probably 'Mediation of Pattern' rather than 'Mediation of Geometry'. So, while the focus in this present book is principally on regular patterns (i.e. those entities which depict an identical element repeating at a pre-set distance across the plane), where possible, the use of the words 'ornament' and 'decoration' will be avoided and the term 'embellishment' used instead, as the latter is more suggestive of something which enhances rather than something which lowers the value; the two words, 'ornament' and 'decoration', will however be retained where necessary within citations from other authors.

As highlighted by Durant (1986, pp. 139–161) 'Orientalism' was a major force influencing the development of design in the nineteenth and twentieth centuries across much of Europe. The Alhambra Palace complex in Granada (Spain) was a favourite source of inspiration for the development of regular patterns, as were the textiles made in Kashmir (India) and the architecture of ancient Egypt. Japanese influences were substantial also, for here, asymmetrical yet balanced visual arrangements were the norm (Durant 1986, pp. 162–191). As observed by Durant (1986, p. 193), Darwinism stimulated the mistaken belief that culture and art developed in 'progressive evolutionary' stages, that cultures regarded as 'primitive' were at a low stage of evolution, and that visual art classified as 'primitive' was immature and thus at a lower stage of development than art of European children. Relevant publications were identified by Durant (1986, pp. 193–194).

It is worth mentioning that contrasting views to Darwinism were expressed by Collingwood (1883), whose opinions were probably more in line with typical late-twentieth and early twenty-first century perspectives. By the early-twentieth century, it appeared that the impact on worldwide visual cultures of visual arts from outside both Europe and North America was substantial, and this was stimulated further by the involvement of artists such as Gauguin and Picasso using visual-art examples from these sources as a stimulus for their own developments; the influence on Cubism and Fauvism appears to have been substantial (Durant 1986, p. 197). The development of such thinking may well have been stimulated by the inclusion of plates of Melanesian and Polynesian bark fabric designs in Jones's *The Grammar of Ornament* in 1856 and a plate entitled 'Decoration from Oceania and Central Africa' in Racinet's *Polychromatic Ornament* in 1873.

It is assumed often by non-textile specialists that all forms of design are suited to all techniques of textile manufacture, but this is not the case. In conventional weaving, all figures and shapes are formed through the raising or lowering of warp or weft threads. In knitting, figures and shapes are created by loops. In the wider area of design application to textiles, at least among mechanised or semi-mechanised techniques, printing offers probably the most extensive graphic capacity, but even here there are various restrictions imposed by the process. In flat-screen printing, for example, difficulties are presented when attempting to print continuous straight lines in a warp-ways direction (due to the mechanical impossibility of aligning successive screens and the lines from one screen to the next with absolute precision). With all sorts of screen printing, the viscosity of the print paste and the porosity of the screen are determining factors on the width of the line that can be printed. The viscosity of the print paste is of importance; if too thin, it tends to bleed (i.e. spread), resulting in a slightly thicker line than desired, and if too stiff, it tends to block the screen, resulting in parts of the design not being printed.

With the various considerations mentioned above in mind, the objectives of this chapter are to introduce the topic, not just to those concerned with the design of regular patterns on textiles and similar surfaces, but also to visual artists in general and designers in other areas, with the intention of stimulating an awareness of regular patterns and how the 'rules' associated with these may be imported with fruitful results into other visual areas. A principal focus of the book is on the partition of space and this is a challenge not just for textile designers but also for practitioners in all visual-art-and-design areas.

Bowles and Isaac (2012 [2009]) explained a wide range of procedures of value to both students and experienced practitioners keen to develop knowledge of textile printing in the digital age. The focus on Adobe Photoshop and Illustrator (and similar) software and the use of these as design tools are to be welcomed, as is the step-by-step approach taken. The drawback is of course that in an era of fast-paced technological change, such publications are prone to redundancy as newer versions of a software and/or new modes of doing things become the norm. Writing in the late-twentieth century, Phillips and Bunce (1993, p. 6) observed that regular patterns could be detected in ancient and relatively modern times, culminating 'in the high-tech computer-aided designs of today', a sentiment still of relevance in the early twenty-first century.

The most common organisation of repeating components is the so-called 'block repeat', where a design is organised on a simple square grid. Numerous variations are, however, possible through dropping alternate columns or sliding alternate rows to produce various 'drop' or 'brick' formats. Phillips and Bunce (1993, p. 85) observed the similarities between brick repeats and half-drop repeats and noted that one rotated through ninety degrees produced the other. With brick repeats, there is thus a horizontal emphasis, while with half-drop repeats, there is a vertical emphasis. So, differentiation is simply a matter of orientation.

The term 'counter-change' is used often when the constituents of a regular pattern exchange colour on a regular basis. Phillips and Bunce (1993, p. 151) observed that the term 'counter-change' was applied to designs in which there was a positive/negative effect of some kind, and colours changed positions on a regular basis. Occasionally, the constituents of regular patterns may take up unexpected diagonal effects. Phillips and Bunce (1993, p. 133) observed that this could be alleviated by placing the components in a sateen arrangement (of the type used by weavers) to suggest a random distribution of components.

1.2   ORGANISATION

This present book is divided into seven chapters covering a discussion of the diversity of patterns within the human environment; a recognition, review and discussion of visual fundamentals of relevance across the visual arts and design; an identification of categories of pattern based largely on their structural characteristics, highlighting types of grid used conventionally in the construction of regular patterns and an explanation of the nature of symmetry operations and how these are of relevance to pattern construction, identifying various sources which it seems could be adapted as compositional grids in order to extend the possibilities and add to the designer's repertoire in the production of fresh and original designs.

1.3   INCLINATION

There is the inclination among humans for order, repetition and regularity, for pieces to be identical in shape, size and content and positioned equidistant from each other. It may well have been that repetition of a feature came about through a technical process of some kind such as weaving; stripes could be produced through alternating colours of warp or weft sets of threads and checks through alternate colours of both sets, each action taken in the human quest for variation, on the one hand, yet comfort with regular repetition, on the other.

Occasionally, it seems that even in circumstances where the medium used does not necessitate regular repetition but, instead, gives flexibility on where a mark, marks or areas of colour are placed on a surface, there is still a tendency to create regular patterns. Examples include batik, embroidery, felting and, to some degree, rug weaving. In the case of batik, it can be assumed, occasionally, that there was the desire to imitate some varieties of woven textile where regular repetition was a feature, and some twentieth-century regular batik designs from Indonesia appear to have such origins; but other Indonesian batiks which also show regular repetition of units appear to have no such models. Some embroidered textiles (such as late-Ottoman-period Turkish towels) often depicted motifs recurring with regularity in border form. In twentieth-century Central Asian felts, regular patterns were found frequently, particularly in border form, as was the case also with twentieth-century Oriental rugs (Persian, Caucasian or Turkish). So, it appears in these cases that there was a tendency often for the producer to impose a regular pattern where the technique did not necessarily demand it, and regular-pattern creation was thus a conscious decision. Further to this, even in circumstances where strict repetition had not resulted, there is the occasional tendency for the onlooker to get the impression of a regular pattern. This is the case often with twentieth-century adire fabric from West Africa, where the onlooker gets a first impression often of a regular pattern with repeating units yet, after close examination, realises that, although a regular framework (of rectangles or squares) holds the design together, component motifs within (rectangular or square) unit cells are deliberately different. So, in this case, there appears also to be a tendency for the viewer (user or purchaser) to perceive a regular pattern, even when it is not present.

So, regular patterns are created by producers in circumstances where the technique used could produce alternatives, and there is a tendency also for the onlooker to perceive regular patterns where they do not exist. Regular patterns resonate strongly, it seems, as part of the human condition/psyche. Why? It appears not to be known. Maybe there are conscious and unconscious layers of needs, desires and associations? Unknown these may be, but what is certain is that all known human societies, historically and in modern times, have had what appears to be an irresistible tendency to embellish what they produce. Initially, marks on the surface of an object may have held a symbolism as they reminded the producer or user of some known person, object or event, physical or spiritual; in time, these may have entered the lexicon or catalogue to be applied in future production of similar objects or items, and, with each stage of production over time, the image may have deteriorated in clarity. So, by the early twenty-first century, often the thematic origins of motifs and regular patterns depicted, for example, on ikats, carpets, batiks and embroideries, were not easily recognisable to the cultural outsider. Westphal-Fitch et al. (2012) confirmed that 'humans spontaneously impose order on visual arrays, using various generative rules' and that 'adults can easily recognise patterns of various sorts' and, in particular, 'correctly identify violation of these patterns'. They maintained that humans had a 'strong drive to impose order on random arrays' and that they did so 'without instruction'.

1.4 CATEGORIES

Prior to the explanation of the range of pattern categories, it is worth stressing that there are various crucial questions to be asked and answered at the early stages of designing a textile product. These include the following: What is the anticipated end use? Which specific properties are required of the finished product? Which technique of manufacture (e.g. printing, weaving or knitting) should be used? Which colour palette and dyestuffs should be used? What is the intended size of the motifs and patterns in the anticipated application? It is important that the design process begins with an end use, for this will guide the series of decisions to be made by the designer; different end uses have different requirements, and each end use will demand specific physical and aesthetic properties of the final product. The selection of fibrous raw materials and technique of manufacture is of importance also (though these may be set in advance by the manufacturer); some designs may be suited to one avenue of manufacture but not to another. The non-specialist should note that different techniques, such as printing, weaving and knitting, will each have technical variations and each variation will have its own restrictions, so the designer may need to ensure that resultant designs take these restrictions into account. It is worth taking note of Bier's (2008) view that the processes of 'pattern making in all media rely upon the interaction of craft and technology' and knowledge of this technology is a prerequisite for effective design.

The rendering of a motif or pattern using one technique of manufacture (such as hand-block printing) will result in distinctive characteristics compared with the use of other techniques (e.g. rotary screen printing). Different workshops may offer different conditions (e.g. variations in temperature or viscosity of print paste may affect dye uptake). The colour palette used will crucially determine whether the final design (or design collection) is commercial. When the wrong colours (i.e. colours not acceptable to a given market) are used, the design will not sell. The intended scale of motifs and patterns is often not communicated by student designers, though, again, this will be a crucial determinant of a design's success. Traditionally (until around the mid-1960s) small-scale component motifs (an arbitrary 1–5 centimetres in the longest direction) were typical for textile designs intended for garment use and large-scale motifs (arbitrarily, over 5 centimetres in the longest direction) were used typically for furnishing end uses. This 'rule of thumb' changed subsequently, and, by the late 1960s, many regular patterns did not adhere to this general size rule. The intended size must however be indicated by the designer, as such knowledge is crucial when setting up the processing technology.

Effective design relies on the designer having a well-developed familiarity with the raw materials used and the technique employed. This was recognised in the first half of the twentieth century by Fenn (writing initially in 1930) but is still of relevance in the early twenty-first century. He commented that if 'a design is to be realized in some material, the tools and the process employed are important factors', and the more these are understood by the designer, the 'better and more economical will be the result' (Fenn 1993 [1930], pp. 41–42).

All regular patterns consist of units of repetition (or repeating units), containing often one or more components known as 'motifs', with each unit of repetition of identical size, shape and content and consisting of exact combinations and distributions of lines, dots, textures and colours. In the case of regular all-over patterns, these units of repetition are reproduced by placing exact copies at precise distances vertically and horizontally so that the resultant all-over effect consists of a unit undergoing repetition covering a surface (or two-dimensional plane). In the design of regular patterns, it is common to present these as some form of surface addition, though occasionally this may not be the case. In the textiles context, a regular all-over pattern can either be added to the surface of a previously produced item, as is the case with printing or embroidery, or the pattern can be integral to the structure and is created through the fabrication process itself, as is the case often with weaving or knitting.

The regular repetition of units is governed by various geometrical rules or symmetries. Symmetry acts as a constraining force, on the one hand, but gives systematic order, on the other hand. A constraining force ensures that there are various geometrical restrictions to the underlying structure of regular patterns, and systematic order is the universal characteristic of all regular patterns. In the context of regular all-over patterns, this order is guided often by reference to a compositional framework or grid consisting of a regular distribution of cells created from polygons of various kinds. These systems of repetition form the compositional 'grammar' of all regular patterns. and this grammar forms the wider focus of attention in the present book. As noted previously (in the Preface of this present book), Leborg (2006, p. 5) perceived parallels between the organisation of a language and the organisation of a regular pattern and noted that his intention in his treatise *Visual Grammar* was to define basic elements and describe patterns and processes in terms of the relationship between the individual parts of a visual system. This is partly the intention in the present book too.

Possibly, the fundamental types of regular all-over textile patterns are those which feature spots or stripes or checks; in each case, the effect may be created during or after the fabrication process itself. Spots and spotted arrangements are found commonly in the natural world, and it may well be the case that early attempts to produce regular all-over spot patterns referred to such sources. In nature, however, strict repetition, without change in scale, colour, texture and orientation, is not a feature. Rather, repetition in nature brings with it minor changes across these characteristics; so, spot arrangements from this source do not exhibit the type of strict repetition of an identical unit or motif characteristic of all regular patterns. Regular spot patterns will be given further attention in Chapter 3 of this book.

Like spot arrangements, stripe patterns occur also in nature, though, invariably do not retain the regularity of industrially produced all-over patterns, as the latter exhibit exact repetition of a unit without variation in size, colour, texture and orientation. Regular stripe patterns comprise a series of parallel bars, oriented identically (vertically or, occasionally, horizontally or diagonally) with the distance between bars and their number, texture, colour and orientation identical; like regular spot patterns, infinite variation appears possible. Regular stripe patterns will be given further attention also in Chapter 3 of this book.

Regular all-over check patterns are common worldwide, though it was in Scotland that they achieved renown as textiles attributed to families or regions due to their dominant colours and other check features; such textiles are known as tartans. Tartans are woven textiles, invariably in 100% wool yarns, dyed to a particular colour palette, placed in a specified order in warp and weft directions, and attributed to a family or geographical region due largely to their colour arrangement and the proportions (or sett) of the woven checked textile created. Key literature was identified previously in Hann (2015, pp. 78–80). Regular check patterns will be given further attention in Chapter 3 of this book.

Other categories of regular all-over patterns include regular diapers and regular medallion patterns. In the context of this book, diapers are those patterns with one or more components which tessellate, i.e. where adjacent repeating units or parts of units share sides, so that the whole surface is covered. Day (1999 [1903]) provided several illustrative examples which he referred to as 'diapers', invariably simple designs based on square or triangular grids; the important aspect was that the repeating units in each pattern tessellated. Medallion patterns, on the other hand, consist of motifs (probably held within a known geometrical shape such as a square, circle, oval or hexagon) repeating on a regular basis on a contrasting (in terms of texture or colour) background. Regular diaper and medallion patterns will be given further attention in Chapter 3 of this book.

In his historical review of pattern categories, Justema (1976) classified patterns under various thematic-type headings: animals (depicting mammals, as well as birds, fish, reptiles and insects, real or imagined), enigmas (described by Justema as 'puzzling', 'mysterious' or 'obscure'), figures (where human-like figures were represented), florals (with floral or plant-like forms), geometrics (where known geometric figures and their derivatives were employed), novelties (where unfamiliar motifs were used), scenics (with landscape-type depictions) and textures (where a textural quality was depicted, maybe through the use of a printing technique of some kind). While Justema placed the emphasis on thematic types of pattern, at the same time, he recognised the importance of considering repeat structures (Justema 1976, p. 10) and even went as far as proclaiming that '...all patterns are basically geometrical', referring, it seems, to the geometrical rules imposed by systems of repetition (Justema 1976, p. 91). A much more direct recognition of the geometrical principles governing pattern construction was provided by Proctor who considered patterns under the headings of various networks (or grids): square, brick and half-drop, diamond, triangle, ogee, hexagon, circle and scale (Proctor 1990 [1969]). The emphasis in this present book will follow the lead given by Proctor (1990 [1969]), though the thematic nature, recognisable often in textile patterns, will be acknowledged where relevant. Also, where appropriate, the means of production will be identified (though not explained in any detail).

1.5 LITERATURE

The intention of this section is to identify various categories of literature of value not only to scholars concerned with regular patterns and their analysis but also to those with interests extending into other aspects of the visual arts and design in general. Four groups of literature are identified below.

First, following in the wake of technological developments in colour printing on paper, principally during the nineteenth century, and the establishment of various national museums (often housing items from distant lands or historical periods), compendia such as Jones's (1986 [1856]) *The Grammar of Ornament* (mentioned also in the preface to this present book) and similar treatises by Racinet (1988 [1873]) and Speltz (1988 [1915]) were published, each with an emphasis on full-colour representation and the identification of items with respect to their cultural or historical sources. Numerous patterns were represented in each, often with minimal explanation. Throughout much of the twentieth century, treatises such as these were stocked in design studios worldwide, where the emphasis was not so much on understanding formal scholarly aspects of the item represented, but rather their aesthetic characteristics which, if captured and incorporated within a design for a product, might have led to market success for the relevant company. Further to this, these and similar publications were often of interest to non-specialist readers due largely to their dramatic visual impact; they became known in the late-twentieth century as 'coffee-table books'. So, an important outcome was that Jones's book and the numerous imitations were picked up and used by generations of visual-arts-and-design practitioners and students, not within formal academic intentions, but rather as inspirational sources. While it is accepted that Owen Jones's treatise is among the most intellectually secure publications of its day, it should be recognised nevertheless that probably its most common use since first publication, was as a design source book for designers and visual artists, rather than as a treatise placed at the centre of a debate relating to the nature of the visual arts historically and in different cultural contexts. Importantly, in addition to the reproduction of an impressive range of worldwide visual arts, Jones presented thirty-seven propositions which he believed were the foundation for good design. Also, it is worth commenting that Jones's work helped to make the subject of regular patterns and related visual arts 'respectable' areas of 'academic study' (Durant 1986, p. 16).

Second, there was a group of books focused more on the structural aspects of the natural and man-made worlds, and these had a great influence on the visual-arts-and-design community of the early twentieth-century. Publications included treatises by Cook (1979 [1914]), Thompson (1966 [1917]) and Ghyka (1977 [1946]). Day (1902) reviewed numerous sources of floral and plant embellishments and considered the application of some in several cultural contexts, including ancient Greek, Egyptian, Assyrian, Persian, Roman, Indian and Gothic. Jackson (1913) debated the application of concepts associated with nature and their inclusion in the design and embellishment of objects, dealing with issues such as repetition, contrast, variety, bilateral symmetry and composition. Related work, placing the concepts and associated thinking into the realms of the visual arts and design, included Meyer (1957 [1894]), Kandinsky (1979 [1926]), Hambidge (1967 [1926]) and Wong (1972).

The immense philosophical contributions made by Gombrich (1979), Wittkower (1978), Arnheim (1974 [1954]), Berger (1972), Dondis (1973), Grabar (1992) and many others should be acknowledged in this category also, though their extensive influence into other subject areas should be recognised as well. Among those concerned principally with patterns and related phenomena (from a non-scientific but structural viewpoint), works by Hay (1836), Hulme (1875), Glazier (2002 [1899]), Day (1999 [1903]), Christie (1969 [1910]), Jackson (1913), Speltz (1988 [1915]) and Fenn (1993 [1930]) are probably among the most noteworthy, though there are numerous others; these are of importance as, in addition to making the reader aware of past achievements in pattern construction mainly from a structural viewpoint, the possibility of further original contributions was stimulated. In the latter part of the twentieth century, the contributions made by Justema (1982 [1968], 1976) to the appreciation and understanding of regular patterns should be acknowledged also. So, this second group influenced developments in the understanding of the structural aspects of the visual arts and design, including patterns.

A relatively unknown treatise concerned with the structural aspects of the visual arts is that by Schauermann (1892, pp. 74–106), who considered various polygonal types and classified each according to its 'regularity', 'semi-regularity' and 'irregularity', with each classification depending on the 'degree of symmetry or equality of angles and sides' (Schauermann 1892, p. 83). Referring to triangle types, he observed that the form 'is regular or irregular as determined by the sides being equal or unequal', and when unequal, 'we get the scalene triangle'; when two sides were equal, an 'isosceles triangle' and with three sides equal, a 'trigon' (or equilateral triangle) was the result (Schauermann 1892, pp. 76–77). Similar statements were made for quadrilaterals (covering various types of four-sided figures, including rectangles, lozenges and squares), pentagons, hexagons and octagons, in each case identifying regular, semi-regular and irregular types, with regular types having all sides equal, semi-regular types with sides and angles equal on opposite sides of a central line and irregular forms with no sides or angles expressing equality.

Third, there were numerous compendia, published throughout much of the twentieth century, often exquisitely coloured with examples collected during 'field work' or held in what were referred to as 'ethnographic museums'. Typically, these set out to identify the types of products, where they were produced and by whom, and included an explanation of the techniques of manufacture, raw materials, motifs and patterns. An important source for anthropologists and others seeking an understanding of historical and contemporary world textiles and their patterns was the monographs produced in the *CIBA Review* series (published by Ciba-Geigy, Basel, Switzerland); this series ceased publication in the late-twentieth century.

The publications listed in this third category were of interest for two reasons: first, to scholars who wished to develop a knowledge of the producing culture and felt it necessary to explore techniques of manufacture, motifs, patterns and raw materials used and second, to design-studio managers who felt that products imitating such items would sell in modern mass markets. In the early twenty-first century, numerous treatises (including many from museum sources) illustrating commercially successful designs from the late-nineteenth and twentieth centuries were published; these appear to have been acquired principally by studios and libraries in higher-education institutions, either intended to allow practitioners to imitate past successes or to stimulate future success among students.

Yoshimoto (1993 [1977]) presented an encyclopaedic publication illustrated comprehensively in full colour, with over 1000 striped and checked textile images, many of the items produced using hand-crafted, resist-dyeing techniques. Bosomworth (1995) presented an extensive collection of black-and-white images (with some regular patterns), mostly line drawings and organised according to continents of origin (Europe, Asia, Africa, the Americas and Australasia including Oceania). A broad treatment was presented, with images selected from numerous sources, including architectural details, sculpture, painting, tiles, mosaics, textiles, ceramics, metalwork, furniture, costume, manuscripts, glass and jewellery. These are but two examples of publications from the late-twentieth century, but many others were published also.

Numerous further publications appeared in the early twenty-first century, often including large proportions of illustrative material in colour and often of value to design-studio libraries. It seems appropriate to classify these too within this third category and to identify a few. Savoir (ed.) (2007) presented an extensive collection of pattern illustrations and types as well as an associated CD of digital images. The publication highlighted the use of familiar pattern types in unusual or unexpected settings. Bowles and Isaac (2012 [2009]) focused on the needs of textile designers and provided a book of value to both students and professional practitioners, with useful step-by-step explanations of the application of Adobe Photoshop and Illustrator and a discussion of how such software was of great value to designers of the day. The text guided the reader gently through the digital creation process, with useful information also relating to digital printing and a particularly worthwhile debate on whether hand-craft techniques could play a role alongside digital developments. In a study of Japanese design, Van Roojen (ed.) (2007 [2003]) presented numerous images in black and white, many from stencils (or katagami); a rich visual source was offered with much of value to the practitioner, though in the absence of detailed explanation, the book was of less value to the analyst. Edwards (2009) presented a wide-ranging introduction to pattern in textile design covering key techniques, themes and design types. This is the ideal publication for non-textile specialists to gain an appreciation of techniques and pattern types and should prove especially useful to gallery, museum and auction-house staff keen to gain a reasonably well-informed view and understanding of textile products.

The fourth group of publications aimed at developing the understanding of the nature of the visual arts, in general, and regular patterns, in particular, was led by mathematicians, crystallographers, physicists and other scientists whose concern was with the identification of symmetry characteristics (particularly in motifs and regular patterns). In many instances, the results offered were by-products of the quest to understand further the crystallographic structures of common substances. Important literature includes Weyl (1952), Shubnikov and Koptsik (1974), Grünbaum and Shephard (2016 [1987]), Loeb (1993), Woods (1935a), Woods (1935b), Woods (1935c), Woods (1936), Schattschneider (1978) and Conway et al. (2008). The comprehensive and widespread nature of symmetry in the natural and manufactured worlds was highlighted in the two-volume treatise edited by Hargittai (1986, 1989); together, the two volumes included over one hundred distinct papers. A further sub-area of interest can be suggested here. This is where an obvious inter-disciplinary approach had been taken, which involved the analysis of designs (including motifs and patterns) from different cultural settings and/or historical periods, with respect to their geometrical, or symmetry, characteristics. Largely, such work had origins in the disciplines of anthropology and archaeology, and typically, stepping outside these disciplines, designs were classified with respect to their symmetry characteristics. So, this area of enquiry has involved also the participation of mathematicians, physicists or other scientists and is thus truly interdisciplinary in nature; the work of Washburn and Crowe (1988, 2004), employing both anthropological and mathematical perspectives, is clearly in this interdisciplinary category.

Durant (1986) provided the most comprehensive review of publications concerned with patterns, focused mainly on the nineteenth and first half of the twentieth century. Durant's book is logically divided into chapters dealing with sources and proposed styles. He recognised the preference among many observers to use nature as a source for developments in the visual arts and discussed the views of Pugin, Jones, Ruskin, Lindley, Forbes, Dresser, Redgrave, Hulme, Haeckel and Binet, all important theorists or practitioners of the day (Durant 1986, pp. 26–28). He recognized the importance of the evolution of *Art Nouveau*, as well as the foundation laid for twentieth-century design by notable individuals such as the American architect Louis Sullivan (Durant, 1986, pp. 28–29). Durant presented a well-developed review of the importance of geometrical construction to the evolution of regular patterns and recognized the important contribution made by Hay (1844), Dyce (1854), Bourgoin (1873), Hulme (1875), Redgrave (1876), Ricks (1889), Glazier (2002 [1899]), Jackson (1913), Day (1902, 1999 [1903]), Ross (1907), Alabone (1910), Christie (1969 [1910]) and many others (Durant 1986, pp. 63–84). Durant (1986) produced probably the most comprehensive review of regular patterns and their study, covering the Gothic Revival of the nineteenth century and the associated views of Ruskin, Pugin and other notable individuals; he also discussed late-nineteenth- and early twentieth-century Eclecticism and the various contributions made by Dresser, Crane, Day, Beardsley, Mackintosh and others; also, he gave attention to Orientalism, Japanese influences, and the visual arts of various 'pre-literate' cultures under the heading 'Primitivism', highlighted first by Jones in the reproduction of images from various Melanesian and Polynesian bark fabric designs in 1856. The importance of the British Arts and Crafts Movement was highlighted together with the role played by individuals such as Morris and Voysey and, finally, a discussion was pursued of the evolution of *Art Deco* and the development of modernism and related developments up to around the middle of the twentieth century (Durant 1986).

An 'unpublished' manuscript produced by Henry van de Velde in the early-twentieth century was discussed by Haddad (2003). Though it is not felt that the translated version discussed by Haddad (2003) adds much to the understanding of patterns beyond other publications published before or since, this treatise is nevertheless worthy of mention as it provides a record of interest and scholarly focus at the time.

1.6 SUMMARY

Throughout known history, each historical and cultural era developed its own regular pattern characteristics and preferences. Due to the derogatory undertones acquired during the late-twentieth century with the terms 'ornament' and 'decoration' and their association with unnecessary additions, it has been decided, where possible, not to use these terms. The term 'regular patterns' will be used where regular repetition without scale change is a feature. It was stressed that for designers to work effectively it was necessary to be aware of the characteristics of the processing technology and to recognize that not all forms of design were suited to all manufacturing techniques. The simplest regular-pattern arrangements, namely block, drop and brick repeating structures were introduced, as was the term 'counter-change'. The inclination among humans for order, regularity and repetition was stressed and the nature of some basic regular-pattern categories, including regular spots, stripes and checks, as well as regular diaper and medallion patterns, was introduced. Various categories of literature were identified, with brief mention being made initially of Owen Jones's grand treatise *The Grammar of Ornament* of 1856. Historically, regular patterns were appropriate to their circumstances of use and (to state the obvious) were produced using the technology and skill available at their time of manufacture. The focus in this book, however, is not on the historical or technological aspects of manufacture but rather on regular structures suitable for the composition of patterns. A selection of images from patterns designed by Guido Marchini (1929–2009) is presented in Figures 1.1–1.33.

EXERCISES

1a   Long-Term Exercise Aimed at Improving Drawing Skills

It is recommended that the following exercise is practiced by all visual-arts-and-design students daily throughout their course of study.

Select a single drawing implement (e.g. a graphite pencil, a crayon, a brush and paint or pen and ink). Select a sheet of paper or other flat surface (any colour, any texture or any weight you may prefer). Without the aid of geometrical instruments (e.g. ruler, pair of compasses or set-square), draw the following figures:

- A circle (of any diameter you may wish);
- A square (of any dimensions you may wish);
- An equilateral triangle (of any side length you may wish).

On a daily basis, feel free to change the drawing implement selected and the type of surface used to apply your drawings. A key outcome will be the improvement and development of drawing skills, so this exercise is of relevance to all visual-arts-and-design students, no matter which discipline. Further or alternative figures (such as a rectangle of proportions of 2:1, 3:1 or 4:3) can be introduced into the exercise should tutors wish.

1b   Regular Border and Regular All-Over Patterns

Use combinations of the three figures drawn in Exercise 1a to produce a regular border pattern and a regular all-over pattern, each using only black on a flat white paper surface. Feel free to overlap, intersect or combine copies of the three figures as you may wish. Ensure that you specify an anticipated means of production, raw materials to be used and an end use for your designs. The final designs should have a hand-drawn quality. Feel free to use tracing paper or scanning and computer software of your choice.

1c   Harmony

Arthur Dow in his treatise *Composition. Understanding Line, Notan and Color* (2007 [1920]) mentioned five criteria which he believed should be considered (and resolved where undesirable) when aiming for harmony in a visual composition. These were opposition, transition, subordination, repetition and symmetry. At this stage, you will have some impression of the meaning of these. For each word, create a single image in black on white (or 'off white') paper, which you believe captures its perceived meaning.

1d   Balanced Composition

Cut or tear the five images created in Exercise 1c. From these pieces, select as many or as few as you may wish and create a paper collage which you consider to be a balanced composition within a rectangle to dimensions of your choice.

1e   A Collection of Regular Floral, Animal, Mechanical or Architectural All-Over Patterns

From a museum or website source of your choice, collect a series of up to twenty photographic images depicting non-repeating subject matter under ONE of the following thematic headings: floral, animal, carnival, mechanical or architectural. Once collected, you are required to consider your images carefully and to carry out drawings of each with the objective of capturing the aesthetic characteristics of each image, in terms of line, form, contrast, colour and texture. From these initial studies, you are required to develop designs for a collection of five regular all-over textile patterns, using a colour palette of up to five colours (plus black and white, should you choose to use these), with an anticipated single end use for your collection, a single method of production and fibrous content of your choice. In addition to your collection of regular all-over pattern designs, you are required also to produce a mood/story board (on a single A4-sized sheet of paper) which shows a selection of your source images, developmental drawings, anticipated end use, raw materials, scale and method of manufacture.

REFERENCES

Alabone, E. W. 1910. *Multi-Epicycloidal and Other Geometric Curves*. London: publisher not known.

Arnheim, R. 1974 [1954]. *Art and Visual Perception*. London: Faber and Faber.

Berger, J. 1972. *Ways of Seeing*. London: Penguin Books.

Bier, C. 1994. Ornament and Islamic art, including a review of *The Mediation of Ornament* by Oleg Grabar of 1992. *Middle East Studies Association Bulletin*, 28: 28–30.

Bier, C. 2008. Art and Mithāl: Reading geometry as visual commentary. *Iranian Studies*, 41(4): 491–509.

Bosomworth, D. 1995. *The Encyclopaedia of Patterns and Motifs*. London: Studio Editions.

Bourgoin, J. 1873. *Théorie de l'Ornement*. Paris: Delagrave.

Bowles, M. and C. Isaac. 2012 [2009]. *Digital Textile Design*. London: Laurence King Publishers in association with Central Saint Martin's College of Art and Design.

Christie, A. H. 1969 [1910]. *Pattern Design. An Introduction to the Study of Formal Ornament*. New York: Dover and previously as *Traditional Methods of Pattern Designing*, Oxford: Clarendon Press.

Collingwood, W. G. 1883. *The Philosophy of Ornament*. Orpington (Kent UK): George Allen.

Conway, J. H., H. Burgiel, C. Goodman-Strauss. 2008. *The Symmetries of Things*. Wellesley (Mass.): A. K. Peters Ltd.

Cook, T. A. 1979 [1914]. *The Curves of Life*. New York: Dover and previously London: Constable and Company.

Day, L. F. 1902. *Nature in Ornament*. London: Batsford.

Day, L. F. 1999 [1903]. *Pattern Design*. New York: Dover and previously London: Batsford.

Dondis, D. A. 1973. *A Primer of Visual Literacy*, Cambridge (Mass): MIT Press.

Dow, A. W. 2007 [1920]. *Composition. Understanding Line, Notan and Color.* New York: Dover and, previously, New York: Doubleday Page and Company, under title *Composition: A Series of Exercises in Art Structure for the Use of Students and Teachers*.

Durant, S. 1986. *Ornament. A Survey of Decoration since 1830*. London: Macdonald.

Dyce, W. 1854. *The Drawing Book of the Government School of Design (1842–43).* London: publisher not known.

Edwards, C. 2009. *How to Read Pattern. A Crash Course in Textile Design*. London: Herbert Press.

Fenn, A. 1993 [1930]. *Abstract Design and How to Create It*. New York: Dover and, previously, under title *Abstract Design: A Practical Manual on the Making of Patterns for the Use of Students, Teachers, Designers and Craftsmen*. London: Batsford.

Ghyka, M. 1977 [1946]. *The Geometry of Art and Life*. New York: Dover and previously New York: Sheed and Ward.

Glazier, R. 2002 [1899]. *A Manual of Historic Ornament*. New York: Dover and previously London: Batsford.

Gombrich, E. H. 1979. *The Sense of Order. A Study in the Psychology of Decorative Art*. Oxford: Phaidon.

Grabar, O. 1992. *The Mediation of Ornament*. Washington DC: Princeton University Press.

Grünbaum, B. and G. C. Shephard. 2016 [1987]. *Tilings and Patterns*. New York: Dover and previously New York: W. H. Freeman and Company.

Haddad, E. G. 2003. On Henry van de Velde's Manuscript on Ornament. *Journal of Design History*. 16 (2): 119–138.

Hambidge, J. 1967 [1926]. *The Elements of Dynamic Symmetry*. New York: Dover Publications and previously New York: Brentano's Inc.

Hann, M. 2015. *Stripes, Grids and Checks*. London: Bloomsbury.

Hargittai, I. (ed.). 1986. *Symmetry: Unifying Human Understanding*. New York: Pergamon.

Hargittai, I. (ed.). 1989. *Symmetry 2: Unifying Human Understanding*. New York: Pergamon.

Hay, D. R. 1836. *The Laws of Harmonious Colouring*. Edinburgh: publisher not known.

Hay, D. R. 1844. *An Essay on Ornamental Design*. London: D. Bogue.

Hulme, F. E. 1875. *Principles of Ornamental Art*. London: publisher not known.

Jackson, F. G. 1913. *Lessons on Decorative Design*. London: Chapman and Hall.

Justema, W. 1976. *Pattern. A Historical Panorama*, Boston: New York Graphics Society.

Justema, W. 1982 [1968]. *The Pleasures of Pattern*, New York: Reinhold Publishing Corporation and previously New York: Van Nostrand Reinhold Company.

Jones, O. 1986 [1856]. *The Grammar of Ornament*, London: Omega and previously London: Day and Son.

Kandinsky, W. 1979 [1926]. *Point and Line to Plane*. New York: Dover and previously as *Punkt und Linie zu Fläche*. Weimar: Bauhaus Books.

Leborg, C. 2006. *Visual Grammar*. New York: Princeton Architectural Press.

Loeb, A. 1993. *Concepts and Images*. Boston and Berlin: Birkhäuser.

Meyer, F. S. 1957 [1894]. *Handbook of Ornament: A Grammar of Art, Industrial and Architectural*, New York: Dover and previously New York: Hessling and Spielmayer.

Phillips, P. and G. Bunce. 1993. *Repeat Patterns*. London: Thames and Hudson.

Proctor, R. M. 1990 [1969]. *Principles of Pattern Design*. New York: Dover and previously as *The Principles of Pattern: For Craftsmen and Designers*, New York: Van Nostrand Reinhold Company.

Racinet, A. 1988 [1873]. *The Encyclopaedia of Ornament*, London: Studio Editions and previously as *Polychromatic Ornament*, London: Henry Sotheran.

Redgrave, R. 1876. *Manual of Design*. London: Chapman and Hall.

Ricks, G. 1889. *Hand and Eye Training*. London: publisher not known.

Ross, D. W. 1907. *A Theory of Pure Design, Harmony, Balance, Rhythm.* Boston: publisher not known.

Savoir, Lou Andrea. (ed.) 2007. *Pattern Design. Applications and Variations.* Beverly (MA): Rockport Publishers.

Schattschneider, D. 1978. The plane symmetry groups. Their recognition and notation. *American Mathematical Monthly*, 85 (no.6): 439–450.

Schauermann, F. L. 1892. *Theory and Analysis of Ornament.* London: Sampson Low, Marston and Company.

Shubnikov, A. V. and V. A. Koptsik. 1974. *Symmetry in Science and Art.* New York: Plenum Press.

Speltz, A. 1988 [1915]. *The History of Ornament.* New York: Portland House and previously as *Das farbige Ornament aller historischen Stile.* Leipzig: A. Schumann's Verlag.

Thompson, D'Arcy. W. 1966 [1917]. *On Growth and Form.* Cambridge: Cambridge University Press.

Trilling, J. 2003. *Ornament: A Modern Perspective.* Seattle: University of Washington Press.

Van Roojen, P. (ed.). 2007 [2003]. *Japanese Patterns.* Amsterdam: The Pepin Press.

Washburn, D. and D. W. Crowe. 1988. *Symmetries of Culture: Theory and Practice of Plane Pattern Analysis.* Seattle and London: University of Washington Press.

Washburn, D. and D. W. Crowe (eds). 2004. *Symmetry Comes of Age: The Role of Pattern in Culture.* Seattle and London: University of Washington Press.

Westphal-Fitch, G., L. Huber, J. C. Gómez and W. T. Fitch. 2012. Production and perception rules underlying visual patterns: Effects of symmetry and hierarchy. *Philosophical Transactions: Biological Sciences*, 367 (1598): 2007–2022.

Weyl, H. 1952. *Symmetry.* Princeton (NJ): Princeton University Press.

Wittkower, R. 1978. *Idea and Image.* London: Thames and Hudson.

Wong, W. 1972. *Principles of Two-Dimensional Design.* New York: Van Nostrand Reinhold.

Woods, H. J. 1935a. The geometrical basis of pattern design. Part 1: Point and line symmetry in simple figures and borders. *Journal of the Textile Institute Transactions*, 26: T197–T210.

Woods, H. J. 1935b. The geometrical basis of pattern design. Part 2: Nets and sateens. *Journal of the Textile Institute Transactions*, 26: T293–T308.

Woods, H. J. 1935c. The geometrical basis of pattern design. Part 3: Geometrical symmetry in plane patterns. *Journal of the Textile Institute Transactions*, 26: T341–T357.

Woods, H. J. 1936. The geometrical basis of pattern design. Part 4: Counterchange symmetry in plane patterns. *Journal of the Textile Institute Transactions*, 27: T305–T320.

Yoshimoto, K. 1993 [1977]. *Traditional Japanese Small Motif. Textile Design I.* Singapore: Page One Publishing and previously as *Textile Design in Japan: Japanese Small Motif.* Tokyo: Graphic-sha Publishing Co.

Figure 1.1–Figure 1.33

Various regular all-over patterns by the Italian designer Guido Marchini (1929–2009) are depicted in Figures 1.1–1.33. It should be noted that the original full-colour versions of these were hand-painted with gouache on paper, in the pre-digital age when much of Marchini's design work was completed. Strict line-for-line and dot-for-dot representation was not required as highly skilled design technicians could readily turn such designs into formal repeats in readiness for production. These were designs for printed textiles, the vast majority for furnishing end uses.

Figure 1.1

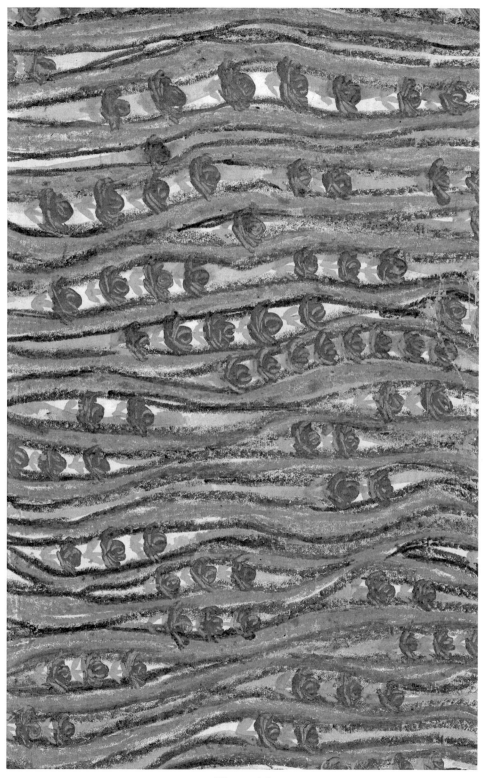

Figure 1.2

Figure 1.3

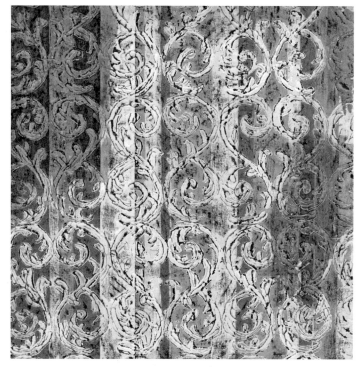

Figure 1.4

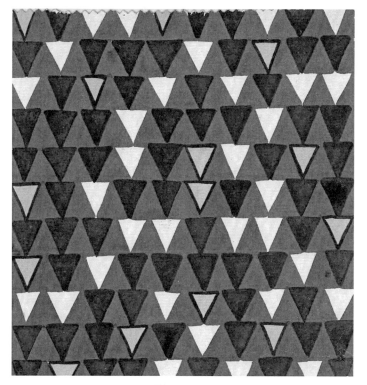

Figure 1.5

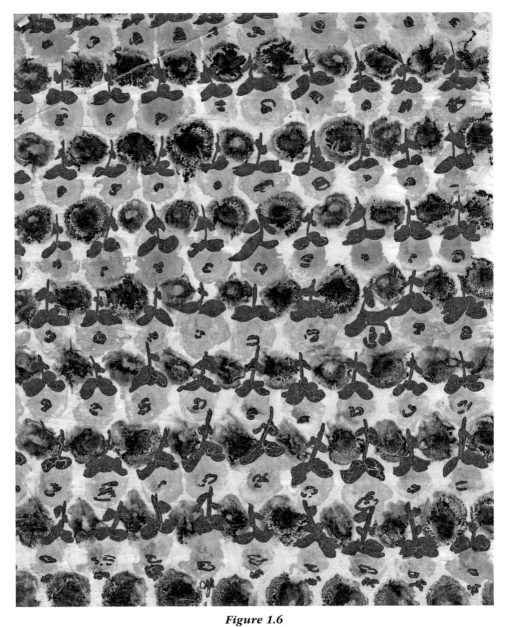

Figure 1.6

Figure 1.7

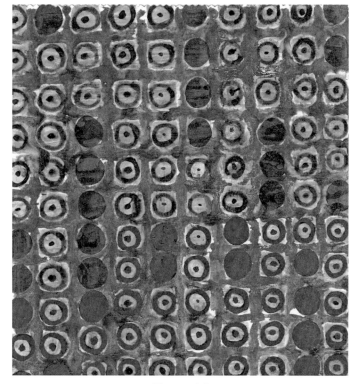

Figure 1.8

Figure 1.9

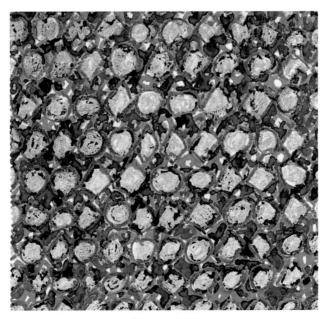

Figure 1.10

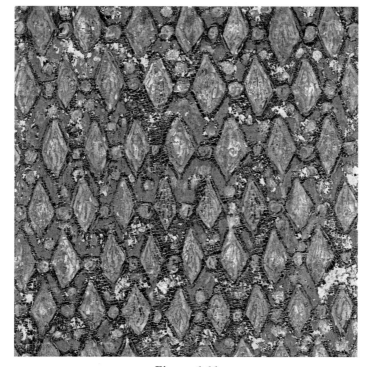

Figure 1.11

Figure 1.12

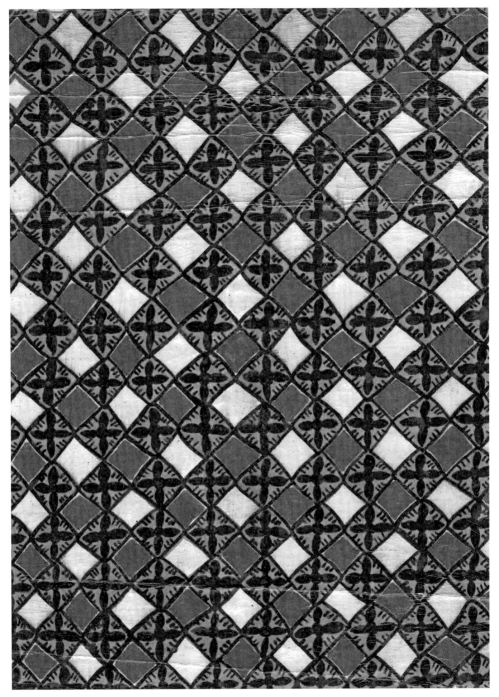

Figure 1.13

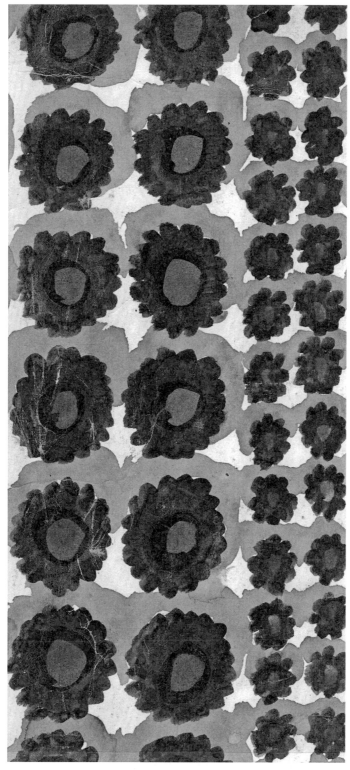

Figure 1.14

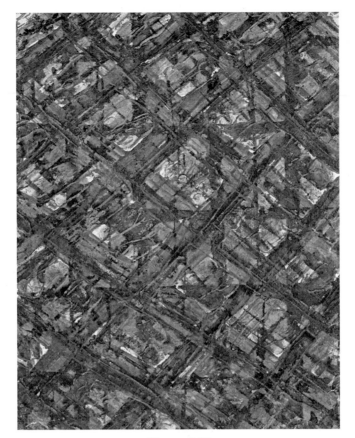

Figure 1.15

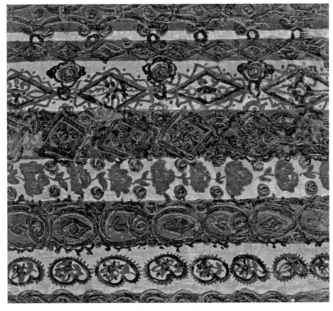

Figure 1.16

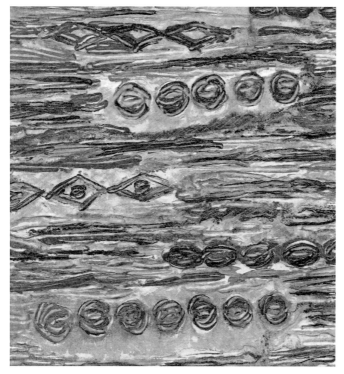

Figure 1.17

Figure 1.18

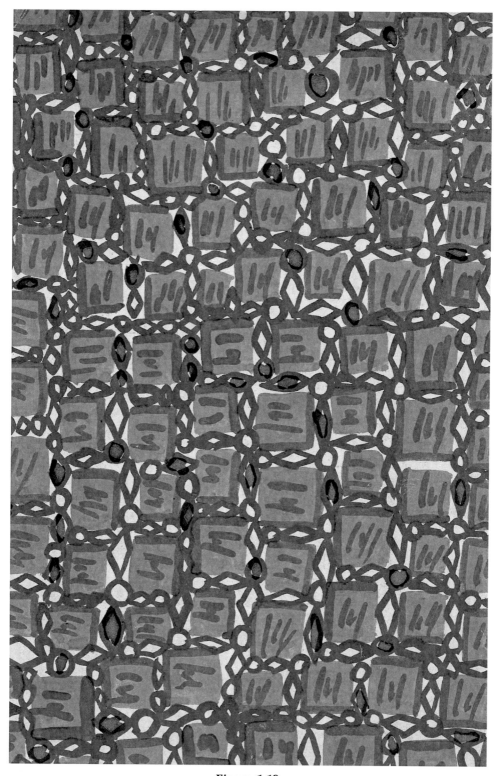

Figure 1.19

Figure 1.20

Figure 1.21

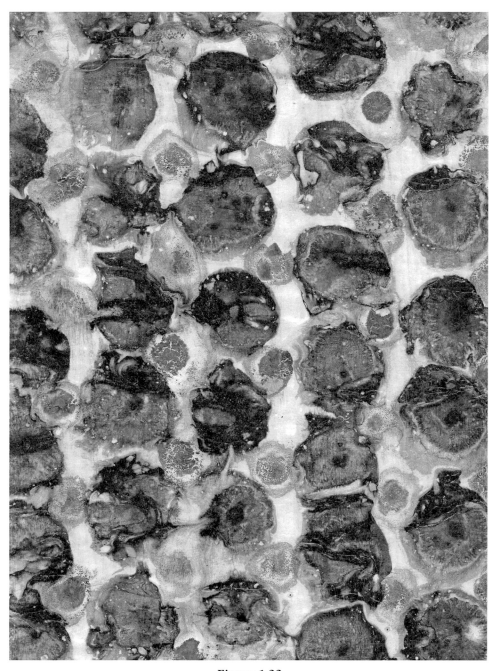

Figure 1.22

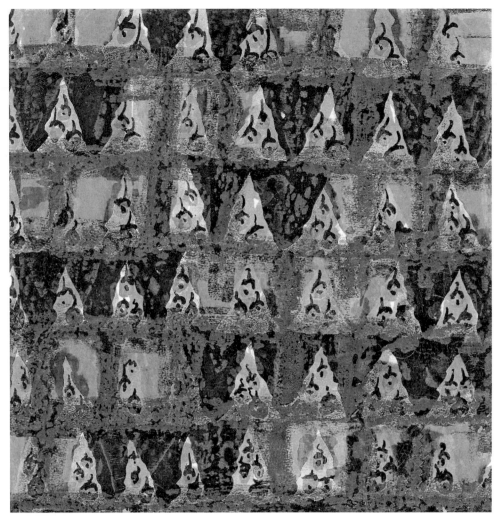

Figure 1.23

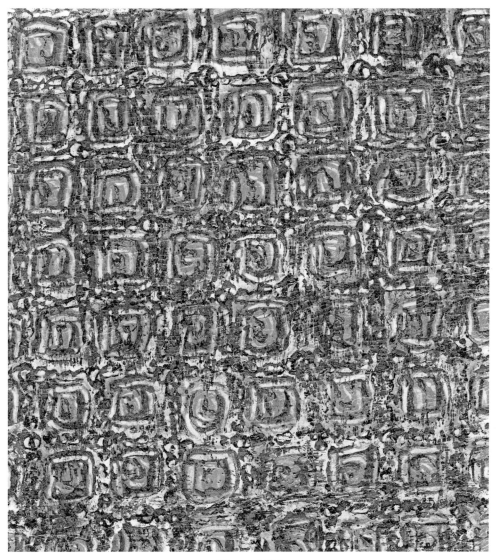

Figure 1.24

Figure 1.25

Figure 1.26

Figure 1.27

Figure 1.28

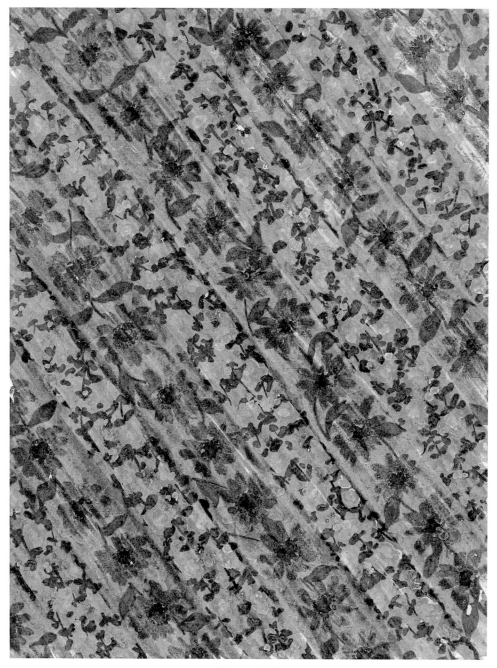

Figure 1.29

Figure 1.30

Figure 1.31

Figure 1.32

Figure 1.33

CHAPTER 2

FUNDAMENTALS

2.1 INTRODUCTION

All areas of visual arts and design have components which practitioners must bring together in ways which are aesthetically pleasing. Prior to the development of a collection of designs, such as textile patterns, designers engage in visual research and consider also the technique of manufacture, the colour palette to be used, the textures to be achieved and the intended end use. A necessity for this preliminary outlook was acknowledged by Steed and Stevenson (2012), who provided a particularly well-focused review, including case studies and interviews with successful practitioners, discussions relating to colours, textures and regular patterns and covered aspects of the development of a commercial collection of textile designs; this publication would be of immense value to both teachers and students, as well as experienced practitioners. Meanwhile, Edwards (2009) provided a well-illustrated guide and reviewed a comprehensive range of textile types from historical and relatively modern periods, particularly the eighteenth, nineteenth and twentieth centuries, referring to carpet types, printing, weaving, lace and tapestry textiles; all in all, this is a treatise of value to both the specialist and non-specialist and would be of great use to museum personnel wishing to develop textile knowledge, as well as to specialists in other visual-arts disciplines who wished to acquire some knowledge of textiles and their manufacture. What is important across all design disciplines is that resultant work should be considered appealing not just to its creator but also to potential users. This wider appeal is based on several factors, including colour palette, texture and the physical arrangement of the component parts; this last aspect is the concern here. The component parts are often termed 'elements', and the identification and organisation of these will be discussed in this chapter. Various regular polygons and their derivatives, and how these can be divided or combined to provide frameworks to aid visual composition, are considered also. Two forms of compositional grid are introduced: first, those grids which are based on a single, stand-alone, geometric figure and its divisions and, second, more substantial arrangements consisting of series of regular polygons arranged side-by-side in net formation across the plane.

Elements such as points, lines, textures and colours are often explained well in relevant texts, but only rarely is there an explanation which truly may be of value to practitioners. The truth is that some or all the elements may feature in a regular pattern and its repeating unit, and practitioners do not bring these elements together following some sort of advance plan. Rather, the practitioner is guided by more elusive criteria: instinct, experience, creativity, skill and knowledge together with an awareness of whatever resources seem appropriate and are available at the time. So, much of what is presented in this book is aimed at informing, and precise procedural rules (akin to painting-by-numbers) are not possible. The designer can be informed by this book and, assuming knowledge of the potential offered by the relevant processing technology, can produce something of undoubted value.

In the creation of regular patterns, it is proposed here that geometry is of two distinct types: organisational geometry and content geometry. Organisational geometry underpins the process of regular repetition and content geometry occurs when known geometric shapes (typically, circles, squares or other regular polygons) feature in the finished form of the regular pattern; organisational geometry is the principal focus within this book. At this stage, it is also worth stating that student designers must realise that complications and lack of clarity should be avoided and that the viewer should never be left confused. So, the use of grids can help with understandable organisation. This calls to mind Puhalla's (2011, p. 8) observation that the principles of visual organisation 'are anchored in the idea that the simplest interpretations of forms and space are preferred'. So, understanding, simplicity, clarity and lack of confusion should be the aims in all forms of visual communication. Puhalla's (2011) publication provided a well-rounded explanation of factors governing good composition and paid attention to the elements of point, line, plane and volume and what he referred to as the 'attributes' of 'shape, size, color and texture' (Puhalla 2011, p. 7). Various fundamental elements are considered briefly in the next section of this chapter.

2.2   POINT, LINE, PLANE, STRUCTURE AND FORM

A 'point' (occasionally referred to as a 'dot') is listed in textbooks invariably as the first, or most basic, element in visual arts and design. It is indeed of importance, for it is the initiator of all that follows, and it is from here that everything else in the visual arts and design will spring. In the conceptual sense, a point is considered to have no dimensions, but is given these often to permit visibility (represented invariably in circular format by ink on paper). Points of different colours can be brought together and combined (or mixed) visually to produce a surface of a different colour, a possibility exploited by pointillist painters in the twentieth century. Also, with various forms of newspaper and magazine printing, the density of points printed on a white surface can create areas of dark and light; so, a high density of black points can yield a surface that appears black and a lower density of black points, a surface that appears grey. Puhalla observed that a point 'is the simplest component of the elements' and that in terms of mathematics, 'a point has no dimensions', so when represented, it must be exceedingly small in comparison with the total image area (Puhalla 2011, p. 28).

The term 'spot' may be used alternatively, especially where dimensions are clearly more substantial and the entity is deemed by the unaided human eye to have length and width. Points are used often as location markers in visual geometry and are combined also with other points in a wide range of visual compositions. A further term, 'polka dot', is used to refer to a regular pattern consisting of circles of the same size printed against a contrasting background. Traditionally, regular polka-dot patterns were associated with clothing worn by Spanish Flamenco dancers but, in the twentieth century, they became more closely associated with children's wear across much of Europe.

Kandinsky (1979 [1926], p. 25) offered explanations of several fundamental elements, maintaining that the 'geometric' point was an 'invisible thing' and, considered in terms of 'substance', it equalled 'zero'. In practicality, in its material form, the point is thought of as a small disc, but is recognisable also if it assumes a variety of shapes; it can be jagged, pointed, smooth or conform to any irregular shape, or take the form of any known regular polygon, so long as it is regarded as 'small' (a relative characteristic, depending on its surroundings). Kandinsky (1979 [1926], p. 50) considered the 'special' question of texture when exploring the nature of point across several techniques, including etching, woodcut and lithography. He believed that the characteristics of a point were determined by the nature of the tool that created it and the characteristics of the surface it was placed on, thus suggesting that a point's characteristics would vary depending on the medium and surface being used. Combining crayon, or pen and ink, or graphic pencil, or charcoal or acrylic paint and brush (or stick) with tissue paper, or linen canvas, or sand paper or newspaper will, with each combination, create a point with a different visual texture.

In the pecking order of fundamental elements, 'line' is presented generally after 'point' and is a one-dimensional entity created by the path of a moving point (Poulin 2011, p. 20). Kandinsky (1979 [1926], p. 57) considered that 'the geometric line' was 'an invisible thing' created 'by movement' and was thus a 'leap out of the static into the dynamic'. Although lines are considered to have one dimension only, they are represented often with width (to aid visibility). Lines have direction and may be curved, dotted or zig-zagged, but are considered invariably to be straight unless specified otherwise. Like points, lines may be in different colours and textures and, depending on the medium and surface used, can have a multitude of characteristics. Lines may be used to give energy (Hann 2012, p. 15) or direction (Oei and De Kegel 2002, p. 77). All geometric shapes are constructed from lines (often with points used as location markers). Everything that can be seen has a shape (Wong 1972, p. 7), which varies for most three-dimensional objects depending on the position of the viewer. Meanwhile, shape remains constant for all two-dimensional entities and is considered composed of one or more lines (or an 'outline') around its outer circumference. Referring to a line, Kandinsky (1979 [1926], p. 57) declared that the 'geometric line is an invisible thing. It is the track made by the moving point; that is, its product'. Puhalla considered a line to be best characterised as 'the path of a moving point' which 'transitions and transforms figuratively into line' (Puhalla 2011, p. 32).

Puhalla observed that shapes were 'self-contained outlines or surfaces' defined by 'regular polygons or variable-sided polygons and closed curved configurations' (Puhalla 2011, p. 42). A plane is a single closed shape with its outline created from one or more lines placed, invariably, on a flat surface. It has position, direction, length and breadth (but no thickness). In two-dimensional design, each plane has an independent existence (as is the case often with, e.g. a newspaper page, a poster or a roll of printed wallpaper), while in three-dimensional design, an inter-relationship to other planes within the design appears to be the norm, as with walls in a room, the pattern components of a jacket (cut initially, in the plane) or the outer panels of an automobile.

The term 'structure' may be used to refer to the underlying order which governs the relationship between design elements (Wong 1972, p. 23). Structure is the underlying discipline of all regular patterns. It predetermines the relationship between the constituents and is imposed often by a framework known as a 'grid' (grids and related structures are introduced briefly in Section 2.5 and considered, subsequently, in Chapters 4 and 6). Structures are evident in the three-dimensional built environment, where the term is used often to refer to a finished object or construction rather than an underlying framework. In the context of this book, the term 'structure' will be used to refer to an arrangement or organisation of components which underpins or acts as a skeleton for an exterior entity known as a 'form'. Structure is probably the most important compositional property, in both two- and three-dimensional designs. It is a determinant of the distribution of parts (shapes, motifs, colours and textures). Structure has a strong influence on whether a design is successful and looks complete and finished (with each part playing a role and each in the correct place). Detailed knowledge of structural possibilities, based on grids or similar frameworks, will not on its own lead to a highly successful and fully resolved design, but it will help to inform the designer by offering an organisational framework on which to place parts.

The term 'structure' can be found in use in the realms of music, chemistry, biology, mathematics, history, literature, philosophy and physics, and commonly, in engineering and building, it refers to the shape taken by a finished construction. In the context of this present book, however, the term will be used to refer to an arrangement or organisation of components which underpin or act as a skeleton to an exterior entity known as a 'form'. The term 'form' refers to the overall shape created by an amalgamation of points, lines and planes and can be used to refer either to a two-dimensional or three-dimensional composition. Poulin considered forms to consist of surfaces and edges which could be split further into shapes, lines and points (Poulin 2011, p. 42). Importantly, a form is the final visual statement, which is underpinned by a structure and, when considering two-dimensional phenomena, this invariably relates in some way to one or more known geometrical figures. The intention here is to ensure that the reader has a familiarity with the range of geometric figures and their derivative constructions which may play a part in the physical composition of regular patterns or other visual-arts constructions. It is the contention here that the aesthetic performance of a form is determined by the nature of its underpinning structure. This is true also in the context of regular patterns which are underpinned invariably by a grid structure which allows one or more repeating units to undergo regular repetition. The underpinning structure may or may not be apparent in the finished form.

Puhalla presented various comments relating to colours and the three important attributes of 'hue, value, and chroma' (Puhalla 2011, p. 62), observing that the hue 'is the generic family name' of the colour, that value was 'characterised by its relative lightness or darkness' and that chroma had two distinguishing characteristics, namely 'saturation and brightness' (Puhalla 2011, p. 62). As the focus in this book is on structure and form in regular patterns, and only limited reference is made to colour, the reader should not assume that colour is unimportant. To the contrary, it is worth noting that a mediocre collection of regular patterns, 'coloured' with a commercial colour palette may achieve market success, while a relatively better resolved collection of regular patterns, if 'coloured' with a non-commercial colour palette, may have little, if any, market success. So, although the emphasis in this present book is on structure in regular patterns, the reader should not assume that colour is an unimportant issue but, rather, should view colour and its understanding as a crucial further challenge after mastering the contents and addressing the difficulties of the accompanying exercises to the present book. It should be noted at this stage that there are several occasions where a regular pattern comprises simply a single foreground colour and a single background colour. On other occasions, a collection of regular patterns may rely on a colour palette of around six colours or more, and each regular pattern within the collection will be coloured in different ways using the given palette to give 'colourways' for each pattern. So, it is best to realise that often each regular pattern is comprised of several colours or layers between what could be referred to as 'front foreground' and 'back background'. As a rule, costs of manufacture (e.g. of a printed textile) will increase as the number of colours increases. So, often experienced designers will attempt to ensure that the intended number of colours in any design is kept to a minimum.

2.3   *REGULAR POLYGONS*

A regular polygon is a closed, two-dimensional figure bounded by equal-length, straight lines (Rich 1963, p. 7, Hann 2012, p. 5) and with equal interior angles. The number of regular polygons is (in principle) infinite. Where sides are not of equal length, and interior angles differ, then the word 'regular' is not used. Triangles of various kinds, quadrilaterals (such as squares and other four-sided figures such as rectangles), pentagons and hexagons, etc. are examples of polygons. All polygons are formed by sides (or lines), vertices (the points where adjacent sides meet) and angles (between adjacent sides). Often, polygons, either on their own or in combination, act as compositional frameworks underpinning and guiding the placement of components of a visual statement. Occasionally the word 'convex' is added to denote a polygon with all interior angles less than 180° and all vertices pointing outwards from the interior of the figure. The concern here is with regular convex polygons of the types used as unit cells for tilings or grids. Various regular convex polygons (from a three-sided equilateral triangle to a regular ten-sided decagon) are presented in Figure 2.1. As the number of sides (or number of connecting lines) increases, the polygon's shape becomes progressively like a circle (Jacobs 1974, p. 499).

The vast bulk of readers will be aware that a square is a four-sided regular polygon, with equal length sides and interior angles of 90° at each vertex. Further to this, a square has four-fold rotation symmetry (i.e. it can be rotated around its centre and maps onto itself in four stages to reach its initial starting point) and four-fold reflection symmetry also (involving four two-sided reflection axes, one located along each central line between opposite sides and two as diagonals between opposite angles). Explanations of symmetry characteristics and their relationship to regular patterns were covered previously in Hann (2012, pp. 72–105) and are dealt with in Chapter 5 of this present book. Squares have been used in the construction of numerous further figures employed commonly in the visual arts, both historically and in relatively modern times. Often these constructions retain rotational symmetry characteristics. A selection of constructions is identified and explained below.

The square is static and at rest when placed on one of its sides. Height and width are equal. Familiarity with the square allows most viewers to consider, imagine or visualise what Puhalla referred to as the five specific 'structural forces' including the vertical axis, the horizontal axis, the diagonal (from upper right to lower left or upper left to lower right), the centre point and a circle drawn inside the square (Puhalla 2011, p. 12). The centre point of a square is indeed easily visualised and, when placed off centre, this is easily detected. Puhalla commented that a 'dot placed slightly off centre will appear misplaced or inaccurately positioned' (Puhalla 2011, p. 28). Also, it should be noted that a point at centre is at harmony and when off centre is in disharmony within the square. A rectangle composed of two adjacent squares is probably the most common rectangle, with an implied centre point identified where diagonals cross. Numerous compositional types are possible in this format (with key components centred, placed diagonally or associated with circles, parts of circles or other regular geometrical figures).

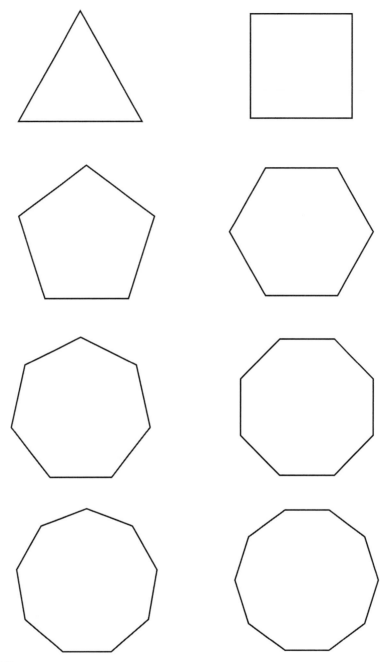

Figure 2.1
Various regular convex polygons (drawing by Chaoran Wang).

A square-based construction which gained much attention in the early twenty-first century was the Brunes star, named after Tons Brunes (1967), an engineer from Denmark who believed that the construction was of potential value to visual artists and designers. Also known as the 'starcut' figure or diagram, this is simply a square, with lines connecting the mid-way points of opposite sides, and also with diagonals from corner to corner and two further half diagonals from each internal angle to the centre points of opposite sides (Figure 2.2). Stewart presented a fascinating treatise and claimed to have found 'evidence' for the use of the construction in 'Chinese, Vedic, Mesopotamian, Egyptian, Greek, Judaic, Sufic, [Italian] Renaissance and contemporary sources' (Stewart 2009, p. 259) and in the 'ground plans visible in the ruins of Jericho, Catal Huyuk and Mohenjo Daro – all more than seven thousand years old' (Stewart 2009, pp. 260–261). A discussion of potential use was given previously by the present author (Hann 2012, pp. 42–46).

A diagonal in a square of sides equal to one unit can generate two 1:$\sqrt{2}$ right-angled triangles (Figure 2.3). The proportion 1:$\sqrt{2}$ is used commonly in the visual arts (Hiscock 2007, p. 201). A square set at 45° inside a larger square (alternatives shown in Figure 2.4) is known as the *ad quadratum* (Hiscock 2007, p. 187). This type of division creates a series of right-angled isosceles triangles with inner angles of 45°, 45° and 90°, with each triangle featuring a side proportion of 1:$\sqrt{2}$ (Hiscock 2007, p. 187). Two equal-size squares can create an eight-point star, by simply placing one square on top of the other (initially with sides and angles coinciding) and rotating it by 45° (Figure 2.5). Also, the overlapped area of the two squares creates a regular octagon; in recent research conducted by Wang (2017), this octagon shape was found to be a common construction in European medieval cathedral floor plans.

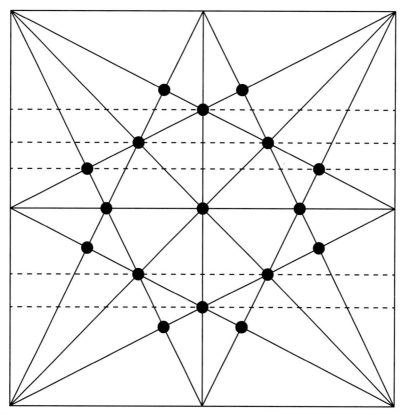

Figure 2.2
The Brunes star with divisions (drawing by Chaoran Wang).

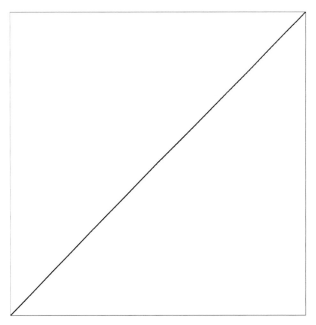

Figure 2.3
A diagonal in a square creating a 1:√2 right-angled triangle (drawing by Chaoran Wang).

The term 'sacred cut' refers to a construction of a square within a square, using four arcs, each centred on one of the four corners of the first square, with the half diagonal length of that square as the radius in each case. The four arcs meet as shown in Figure 2.6 (Watts and Watts 1992, p. 309). When four lines are drawn, which touch the arcs while parallel to the square edges, a nine-unit grid with an inner 'sacred-cut' square results (Figure 2.7) (Wightman 1997, p. 68). Successive inner-square constructions are possible,

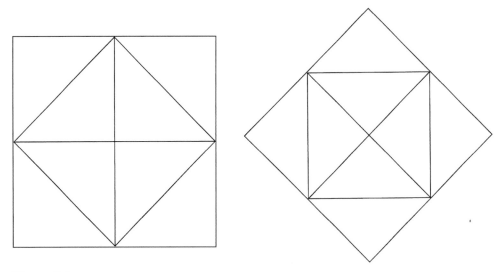

Figure 2.4
Squares set within larger squares (drawing by Chaoran Wang).

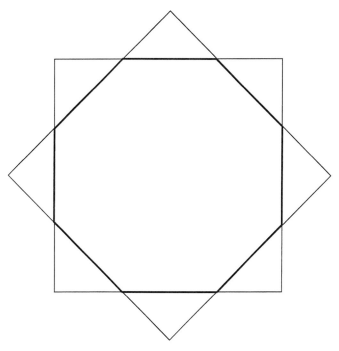

Figure 2.5
A square superimposed on a square (drawing by Chaoran Wang).

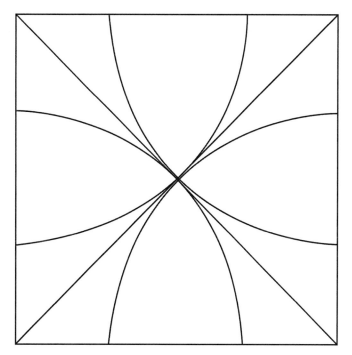

Figure 2.6
Four arcs meeting within a square (drawing by Chaoran Wang).

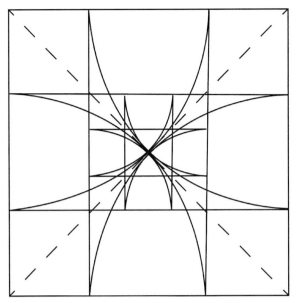

Figure 2.7
A nine-unit grid with inner 'sacred-cut' square (drawing by Chaoran Wang).

thus creating a square within a square within a square, etc. The sacred-cut process can be reversed, with the inner square defining the next largest outer square (Watts and Watts 1987, p. 271) (Figure 2.8). Sacred-cut constructions have been used in Christian cathedrals, Buddhist temples and in Islamic designs of various kinds (Dabbour 2012, p. 38).

It is well known that polygons are shapes bounded with straight lines (Hann 2012, p. 15), and regular polygons are special types of polygons, with equal-length sides and

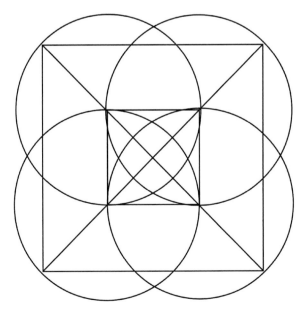

Figure 2.8
A reverse 'sacred-cut' process (drawing by Chaoran Wang).

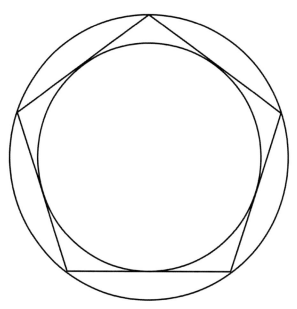

Figure 2.9
A circle within a regular polygon within a circle (drawing by Chaoran Wang).

equal interior angles. All regular polygons are cyclic (Jacobs 1974, p. 499), which means that each can be placed inside a circle with all its vertices touching the circumference of the circle. Meanwhile, a circle can be drawn inside any regular polygon and touch all its sides; an example is shown in Figure 2.9 (Rich 1963, p. 69).

In considering the placement of the parts of a composition, visual artists and designers aim for visual harmony; to achieve this, the constituent parts are placed in relation to each other and the whole in such a way to avoid visual distraction (Birkett and Jurgenson 2001, p. 255). It has been argued often that various rules of proportion associated with known geometrical constructions and based on polygons of various kinds, have evolved, and it has been argued that modern visual artists and designers would reap success if they adhered to the rules of composition associated with these. Constructions which have received the most attention include golden-ratio constructions of various kinds, root rectangles and the Brunes star. Explanation of each is given below as these may be of some relevance to regular pattern construction.

The golden ratio, or 1:1.618, often referred to as φ (phi), and known also as the 'golden section' or 'golden proportion', is believed to yield aesthetic harmony if the constituents of a composition are placed in relation to this ratio (Poulin 2011, p. 223). It was observed previously (Hann 2012, p. 109) that the measure has been the subject of a vast range of enquiry, and a large quantity of literature has resulted. Fanciful claims for its presence have extended to considerations of ancient Greek sculpture, Italian Renaissance paintings, much of nature, the universe and the visual arts and design in general. Indeed, there appears to be the belief that golden-ratio measures can be found in any artefact, structure or construction of repute and it is only a matter of simple discovery to find them; the reader is warned to be vigilant when presented with claims relating to this and related constructions.

A numerical series associated closely with the golden ratio is the Fibonacci series (1, 1, 2, 3, 5, 8, 13, 21…etc.), which displays the characteristic that each number in the series is the sum of the previous two and that the ratio between adjacent numbers is close to that of the golden ratio (Pough 1976, p. 8, Elam 2001, p. 11, Hann 2012, p. 109). That is, for example, 8:13 and 13:21 are each close in value to 1:1.618. The series was named after Leonardo Fibonacci (1175–1250), a mathematician from Pisa, who is reputed to have introduced the use of zero into European counting systems having learnt its use previously in North Africa.

The golden ratio may well be of value to visual artists and designers in deciding where to place components of a composition, but the lack of rigour associated with the discovery of its use to date has, unfortunately, blemished its reputation in the minds of many scholars. A well-focused review was provided by Livio (2002). Relevant constructions are identified below, and the practitioner is invited to experiment. The golden ratio is explained often in terms of line segments, with a straight line divided into two unequal parts, such that the ratio of the shorter segment to the longer segment is the same as that of the longer segment to the whole (Pough 1976, p. 8, Poulin 2011, p. 223, Hann 2012, p. 110). If the longer segment is considered to measure 1 unit, then the shorter segment will measure 0.618 and the whole line 1.618 (Figure 2.10).

A golden-ratio rectangle can be drawn from a square by adding a rectangular component as shown in Figure 2.11. If the sides of the square equal 1 unit each, then the resultant rectangle will have a long-side length of 1.618. Further subdivision and additions can lead to the construction of a golden-ratio spiral within the rectangle (Figure 2.12) or to a rectangle holding a series of squares known by Hambidge (Hambidge 1967 [1926])

Figure 2.10
A golden-section measurement. If the longer segment is considered to measure one unit, then the shorter segment will measure 0.618 of a unit, and the whole line will measure 1.618 units (drawing by Chaoran Wang).

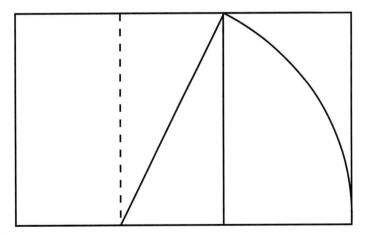

Figure 2.11
A golden-section rectangle can be created from a square by making the additions shown here. If the sides of the square equal one unit each, then the longer side of the resultant rectangle will measure 1.618 units (drawing by Chaoran Wang).

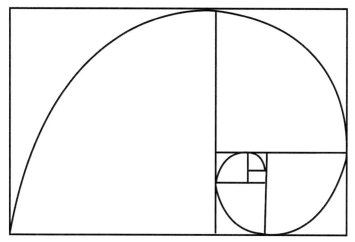

Figure 2.12
A golden-section spiral within a golden-section rectangle (drawing by Chaoran Wang).

as the 'rectangle of the whirling squares' (Figure 2.13), with the sides of each successive square conforming to a Fibonacci series of measures.

A series of root rectangles can be constructed from a square (Figure 2.14). It has been argued previously that root rectangles can provide potentially useful proportions with or without subdivision (Hann 2012, p. 40). In design practice, root rectangles, especially root-two rectangles, are used commonly as compositional aids (in that the dimensions of the composition will conform to the length and breadth ratio of the relevant rectangle). Interestingly, the so-called 'DIN system' of paper sizing is based on root-two rectangle subdivisions (Elam 2001, p. 36). Various paper-sizing standards have existed at different periods in different countries. In the early twenty-first century, the widely accepted paper sizing system (introduced by the International Organisation for Standardisation, or ISO) included the commonly used A4 size, used by most countries outside the Americas (where a different size system is prevalent).

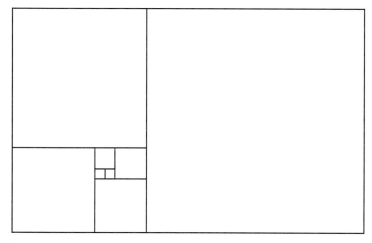

Figure 2.13
The rectangle of the whirling squares (drawing by Chaoran Wang).

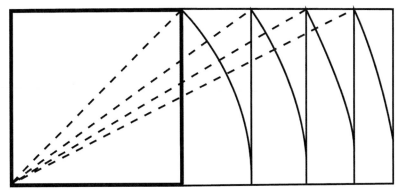

Figure 2.14
A series of root rectangles constructed from a square (drawing by Chaoran Wang).

An explanation of the ISO system was given by Ambrose and Harris (2008, p. 28). Briefly, this is as follows: The largest size, A0 (841 mm by 1189 mm) is twice the size of A1, which is twice the size of A2, which is twice the size of A3, which is twice the size of the commonly used A4, with A5, A6, A7, A8, A9 and A10 each in the series (cut as shown in Figure 2.15).

The relationship between root-three rectangles and the regular hexagon is worth noting. By connecting the opposite vertices of a regular hexagon, three root-three rectangles develop (Edwards 1967 [1950], p. 67). This is illustrated in Figures 2.16 and 2.17. In other words, a regular hexagon can be constructed by a root-three rectangle rotated three times around a central point (Elam 2001, p. 40). A root-four rectangle can be divided into two squares. If two golden-section rectangles are overlapped as shown in Figure 2.18, a rectangle is generated, with a central (overlapped) square.

Rectangles can be divided and then subdivided by their diagonals (Elam 2001, p. 33, Hann 2012, p. 42). This is the case with the Brunes star (above as Figure 2.2), which

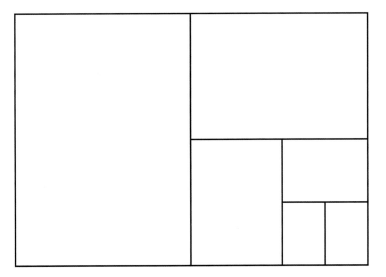

Figure 2.15
An indication of the relative sizing of paper using the ISO system (drawing by Chaoran Wang).

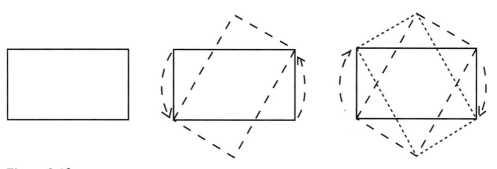

Figure 2.16
Stages in the construction of a regular hexagon from root-three rectangles (drawing by Chaoran Wang).

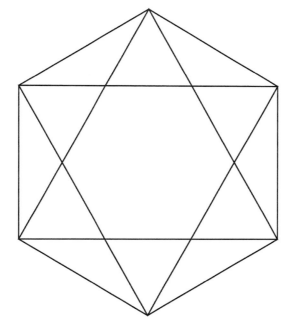

Figure 2.17
A regular hexagon created from three root-three rectangles (drawing by Chaoran Wang).

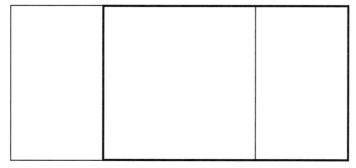

Figure 2.18
Two golden-section rectangles overlapping (drawing by Chaoran Wang).

is first divided into four equal squares, diagonals from corners to opposite corners and two half diagonals from each corner. Points of intersection can then be identified. Overall, the resultant construction shows numerous reflectional and rotational symmetry characteristics and, as such, lends itself well as a compositional figure (Hann 2012, p. 43). There are numerous further constructions derived from the division or further development of the square. Division and reassembly in a different format is an important approach in design. Evidence can be found in pattern design (Thompson 1977, Özdural 2000), religious paintings (Popovitch 1924), graphic design (Behrens 1998, Elam 2004) and church or domestic floor plans (Betts 1993, Wiemer and Wetzel 1994).

Although a single line can divide a square space into desirable proportions, the minimum number of division lines found in common use is two. One of the common methods is to divide a square by its diagonal lines (Figure 2.19). This division can be found commonly in Islamic pattern design (Özdural 2000, p. 175). A proportion of 1:$\sqrt{2}$ exists between the diagonal lines and the square edges, and the square is divided into four equal-size right-angled triangles. These highly modular subdivisions provide flexibility for pattern designs (Özdural 2000, p. 174). Similar division is also found in many medieval-church floor plans (Wang 2017). Wiemer and Wetzel (1994) found the diagonal division of a square when analysing the geometrical structures of the floor plan of the chapel in Ebrach monastery church, a medieval church in Bamberg, Germany. In Islamic pattern design, one method of designing patterns by using square units is to assemble a large figure from five square units (Özdural 2000, p. 175) and rearrange the component parts (including surface designs which may cover these parts). Figure 2.20 shows a square reconstructed from five smaller square units. A larger single square can thus be assembled from these divisions (with four identical

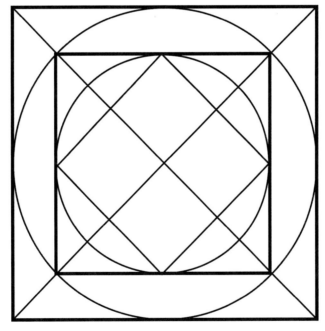

Figure 2.19
A square with diagonals, a circle within a circle and squares within squares (drawing by Chaoran Wang).

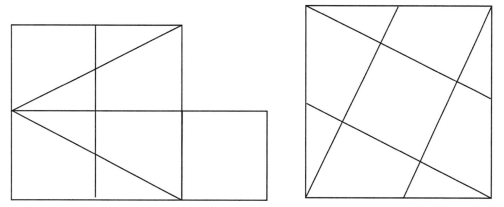

Figure 2.20
A square constructed from five smaller square units (drawing by Chaoran Wang).

right-angled triangles, four identical right-angled trapezoids and a square unit in its centre) (Özdural 2000, p. 175).

In a study of Babylonian geometry, a structure involving 'halving, squaring, addition, subtraction, square roots and multiplying by reciprocals' was found (shown in Figure 2.21) (Bidwell 1986, p. 23). Geometrically, the construction shown is a square within a square with four identical rectangles between the two (Bidwell 1986, p. 23). Any of the four rectangles can be considered to repeat in an anticlockwise (or, alternatively, a clockwise) direction within the larger square. The width of the rectangle is one unit,

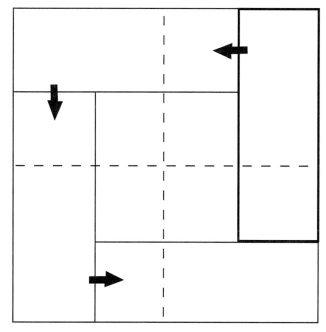

Figure 2.21
A structure involving 'halving, squaring, addition, subtraction, square roots and multiplying by reciprocals', found in Babylonia. (Adapted from Bidwell 1986: 23; drawing by Chaoran Wang.)

while the length is three units; the side of the inner square is two units and the sides of the whole outer square are each four units. Thus, the proportion of 1:2:3:4 is apparent in this division.

Instead of division by straight lines to obtain other constructions, square spaces can be divided by curved lines (or arcs) as well. The most common example of a square space divided by four arcs is in the construction of the so-called 'sacred cut' (explained and illustrated previously in this section).

Watts and Watts (1986, 1987, 1992) focused on garden houses at Ostia, a port city of Rome, and its urban development, dated to around 128 CE, comprised of various forms of living accommodation (mainly apartments), gardens and shops. They found evidence for the use of concentric squares, created by sacred-cut square constructions, which they believed 'guided the design of architects, painters, mosaicists in the production of domestic environments' (Watts and Watts 1987). The Wattses showed that the relevant series of constructional figures was based on the square and its divisions, claiming that this corresponded to 'Vitruvian references to symmetry and commensurability' (Watts and Watts 1987).

Day (1999 [1903], pp. 10–17) presented a series of simple designs, all built from the manipulation of square grid formations, mainly by adjusting angles and orientation, showing various fret, diamond and zig-zag type regular patterns. He showed also how the introduction of a third series of parallel lines on a square grid, passing through and bisecting opposite angles of each square unit cell, could produce a grid of isosceles triangles and, with further manipulation and change in orientation, a grid of equilateral triangles, the basis for numerous design types, could be created (Day 1999 [1903], pp. 18–22). Also, hexagonal-type grids were readily developed from equilateral triangular grids, again the basis for numerous design types. Day showed how further grid types were possible, each building on the simple square grid, sometimes adding a series of parallel lines further apart than the lines of the original square grid, to produce frameworks which included octagon, pentagon and other unit cells (Day 1999 [1903], pp. 23–26). Using circles, various curvilinear versions of numerous grid types were possible (Day 1999 [1903], pp. 27–38). Giving several examples, Day (1999 [1903], pp. 39–52) observed that often there were several methods of attaining the same pattern, rather than one set method.

2.4   CIRCLES AND THEIR DERIVATIVES

In the drawn geometrical sense, a circle is an enclosed shape created by a single line equidistant from a stationary point known as a centre. The distance between any point on the circumference of the circle and its centre is known as the 'radius'. The term 'circle' is used to refer to either the whole disc-shaped figure (including interior and content) or to the boundary or outline of the figure. In the context of this present book, the term 'circle' will be used to refer exclusively to the geometric outline figure only. A series of relevant constructions was given by Lundy (2001 [1998]) and Halmekoski (2014).

A circle is thus an enclosed space, created by a line composed of a series of points in the same plane, with each point in the series positioned at an equal distance to a central stationary point (Rich 1963, p. 68, Jacobs 1974, p. 420, Hann 2012, p. 33). The line created from the set of points is known as a circumference, and a part of the circumference is known as an arc. A line between any point on the circumference and the centre is known as a radius (plural radii). A line joining two points on the circle is a chord. A diameter is a chord which passes through the centre of a circle. A centre, circumference, arc, chord, radii and diameter are shown in Figure 2.22 (adapted from Rich 1963, p. 68).

In Islamic design, the square is often cut by four circles with diameters equal to the side length of the square (Figure 2.23) (Dabbour 2012, p. 383). The square and its inner constructions can be further divided into smaller squares and right-angled triangles (shown previously in Figure 2.4) (Özdural 2000, p. 184).

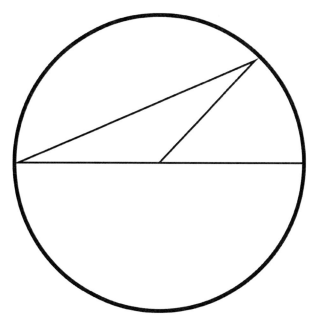

Figure 2.22
A circle showing centre, circumference, arcs, chord, radii and diameter.

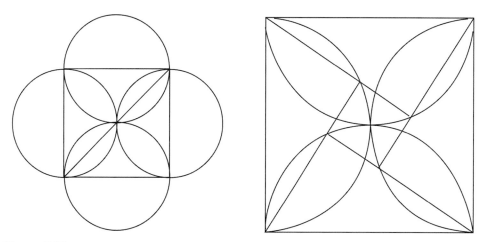

Figure 2.23
A square cut by four circles, each with a diameter equal to the side length of the square (with drawing by Chaoran Wang).

2.5 *DIVISIONS OF SPACE AND COMPOSITIONAL GRIDS*

The grid, as an assembly of lines (Hann 2012, p. 19), has been applied in various design disciplines. Examples can be found in pattern design (Collins 1962), graphic design (Samara 2003, p. 24), architectural design (Chilton 1999, p. 2) and floor-plan structures (Hann 2012, p. 18). The arrangement of text and images on a page follow, invariably, some sort of grid format. Horizontally, a page can be structured according to what is known as 'a base-line grid'. Vertically, this grid is divided into columns and margins (Roberts and Thrift 2002, p. 27), with the column width based on the desired size of type and the number of pieces of type and spaces desired within a line (Roberts and Thrift 2002, p. 18, Samara 2003, p. 27; Müller-Brockmann 2008, p. 31). The number of columns depends on the print format and the size of typeface (Müller-Brockmann 2008, p. 49). Usually, the larger the selected type, the wider the columns required (Müller-Brockmann 2008, p. 31). The aim is to provide the most comfortable reading experience. According to Müller-Brockmann, columns too wide can weary the reader, while columns too narrow force the readers to change lines too rapidly (Müller-Brockmann 2008, p. 31).

The manuscript grid (left of Figure 2.24) is the simplest grid structure used by publishers and is comprised of a dominate rectangle occupying a page (Samara 2003, p. 26). In a book spread, it is usual to create a mirror image, from left to right, allowing also for the margins (Hurlburt 1978, p. 69). The margins are particularly important in defining the appearance of the page (Samara 2003, p. 26).

The column grid (to the centre of Figure 2.24) is of the sort used commonly in newspaper, magazine and book-page design, allowing different information to be arranged in different vertical columns independent of each other (Samara 2003, p. 27). Furthermore, the column structure enables different information to be separated (Samara 2003, p. 27). For example, pictures and text are usually arranged in different columns in newspapers, magazines and webpage designs. Columns can work separately or can be combined to fulfil the requirement of the content (Hurlburt 1978, p. 55).

A modular grid is a column grid with horizontal subdivisions that appear as a matrix of cells (right of Figure 2.24) (Samara 2003, p. 28). Each module is a small space for information (Samara 2003, p. 28), but modules can either work individually or combine to form a larger field to fulfil special needs (Samara 2003, p. 28). For example, large

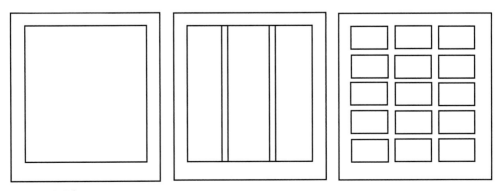

Figure 2.24
Examples of the format of the manuscript grid (left), the column grid (centre) and the modular grid (right), drawing by Chaoran Wang.

pictures, illustrations and tables may occupy several modular spaces in a page (Müller-Brockmann 2008, p. 60). Modular grids are usually divided into 8, 20 or 32 fields (Figure 2.25) in design practice (Müller-Brockmann 2008, p. 90). Grid systems discussed above, however, cannot fulfil all design needs. A grid of a type referred to by Samara (2003, p. 29) as a hierarchical grid may be needed for a project (e.g. poster, billboard or webpage). It appears that such a grid does not follow an exact predetermined format but, rather, is created to meet the specific needs of a project. An example is given in Figure 2.26; this is a hierarchical grid typically used in early twenty-first century webpage design. Samara commented that such a grid was based on an intuitive placement (Samara 2003, p. 29). As with all grid forms, intuition, experience, common sense and creativity are required, and no grid system offers a fail-safe route to superior design.

In addition to printing design practice, grids have also played an important role in other design disciplines. Hann pointed out that, historically, equilateral triangles, squares and hexagons were each of value to design development and could be found

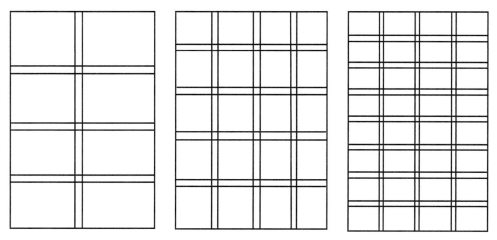

Figure 2.25
Examples of grids with eight, twenty and thirty-two fields (or main divisions).

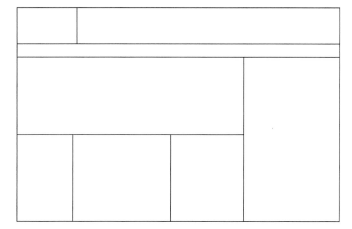

Figure 2.26
An example of a hierarchical grid of the type typical of early twenty-first century webpage design.

often in floor plans or building façade design. Ratios such as 1:2, 2:3, 3:4 or 1:1.618 were each common (Hann 2012, p. 21). Therefore, grid systems have great flexibility and can be divided in different ways to fulfil different design needs. Among all the divisions, modular grids (i.e. grids with numerous interconnected unit cells which can be left alone or combined with their neighbours) appear to have the most flexibility and have been applied in both two-dimensional and three-dimensional designs. In design practice, square and rectangular figures are divided commonly by four straight lines into nine components with two parallel lines placed vertically and two parallel lines placed horizontally, providing a three-by-three compositional structure. Such a structure will provide four intersecting points known as 'optimal focus' points, and key visual elements are usually placed on or near these points (Elam 2004, p. 13). The same division is found commonly in ancient Roman architectural design (Jacobson 1986, p. 16).

The modular division of space can be based on grids (as with the 3 × 3 grids). Also common is the 6 × 6 grid used frequently in design practice. In graphic design, Elam, for example, found that 6 × 6 grids underpinned various poster constructions (Elam 2004, p. 104). The 6 × 6 grid structure was found to exist also in the floor plan of 'Hadrian's Villa' (at Tivoli, Italy), a building plan associated with ancient Rome (Jacobson 1986, p. 77). The geometrical structure identified by Jacobsen is reproduced in Figure 2.27 (with the main components of the structure shown to the right).

In the floor plan of the 'small bath' in Hadrian's Villa, a square and two rhombuses are constructed within a 6 × 6 grid (Figure 2.28) (Jacobson 1986, p. 80).The length of each side of the square is $2\sqrt{2}$ units, while the length of each side of the rhombuses is $\sqrt{5}$ units. Thus, the proportion $1:\sqrt{5}:2\sqrt{2}$ can be found in this construction. Moreover, the 6 × 6 grid can be considered as an elaboration of the 3 × 3 grid. In graphic design, two or more types of grid can be put together to provide interesting layouts. Likewise, in the case of space division, more than one type of division can be combined as well. In the floor plan of Canterbury Cathedral, two identical squares are each divided into 3 × 3 grids and then combined with one at 45° to the other (Wiemer and Wetzel 1994, p. 452, Dudley 2010, p. 46). This construction is illustrated in Figure 2.29. The divided

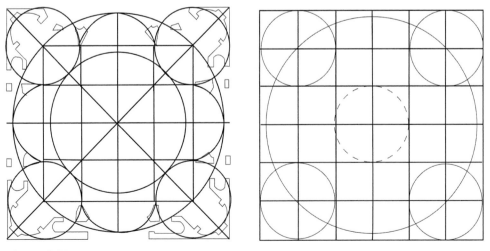

Figure 2.27
The 6 × 6 grid found in the floor plan of Hadrian's Villa.

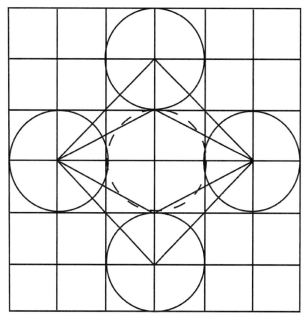

Figure 2.28
Floor plan of the 'small bath' in Hadrian's Villa.

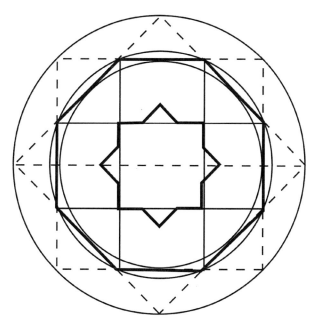

Figure 2.29
Floor plan of Canterbury Cathedral, with two identical squares (each divided into a 3 × 3 grid), and then combined by orienting one by 45° to the other. (Adapted from Dudley, C. J. 2010. Canterbury Cathedral. Aspects of its Sacramental Geometry. Bloomington (IN): Xlibris; Wiemer, W., and G. Wetzel. 1994. Journal of the Society of Architectural Historians, 53 (4): 448–460. Drawing by Chaoran Wang.)

square lines form an octagon (illustrated in bold line in the figure). The inner two squares form an eight-point star.

Elam (2001) reviewed proportion in the manufactured, built and natural environment, referring to various figure types, including golden-section, root-2, -3, -4 and -5 rectangles (and their subdivisions) and analysed a range of posters, consumer products and buildings referring to these. Her procedure was largely to place an image of the poster, object or construction within a rectangle and to divide the rectangle with lines (often diagonals) passing through key components of the design, revealing (she argued) how decisions relating to proportion have adhered to certain known systems. Overall, her analysis appears to suggest that designers used known structures in design development.

Constructions such as brick bonds, traditional Chinese window and door lattices, tilings and stained-glass windows are all means of dividing space and evolved over several centuries. Hann and Moxon (2019) proposed that the structures underpinning these constructions should be adopted and adapted as compositional grids to aid the designer; many of these frameworks exhibit regular repetition and should prove of value to regular pattern designers, as they may offer a further layer of originality. This proposal is extended further in this present book, with the inclusion also of further original illustrative material.

2.6 SUMMARY

This chapter introduced the fundamental structural elements (including points, lines and planes) associated with the visual arts and design in general. Differentiation was made briefly between 'organisational geometry' and 'content geometry', and the necessity for designers to aim for simplicity was stressed. Simplicity is enhanced through planned organisation of constituent elements, and this is most conveniently done using a regular grid. Various fundamental issues were reviewed. For example, line divisions of 2, 3 and 4 were discussed, followed by grid division and an explanation of shape division methods. Proportions, such as 1:2; 1:3; 5:4; $1{:}\sqrt{2}$; $1{:}\sqrt{5}$ and 1:0.618, used commonly in graphic, as well as architectural, design were introduced. A convex polygon, which plays an important role in grid structures, was seen to be a simple figure in which no line segment between two points on the boundary goes 'outside' the polygon. Also, in a convex polygon, all interior angles were seen to be less than 180°. Kandinsky played an important role in reviewing the nature of fundamental elements. Also, it is worth stressing that he recognised that the use of different techniques (i.e. different means of application on different surfaces) would yield different results.

EXERCISES

2a Points

Points are regarded as the most fundamental entities in the visual arts and design, generating lines, polygons, various shapes and related compositions. Create up to twelve images each in black on white paper which you feel encapsulate the meaning of the term 'point'. Cut your completed images (each into say four or more pieces) and reassemble on a white paper background (using as many or as few pieces as you wish) to create what you consider to be a visually resolved image.

2b   Lines

All visual statements are comprised of lines. How these are spaced and combined will determine the success or otherwise of any image. Create up to twelve images which you feel capture the meaning of the term 'line'. As with Exercise 2a, cut each image into at least four pieces. Assemble the cut pieces (again, using as many or as few as you may wish) to create what you consider to be a visually resolved image.

2c   Cut, Combine and Repeat

Consider the outcome of Exercises 2a and 2b. Cut each of these into at least four pieces and reassemble to create what you consider to be an image suited as the repeating component of a regular all-over pattern, with repetition based on a grid form of your choice. Feel free to resize and further edit the repeating unit to make it suited to a stated end use and means of production.

2d   Hand-Drawn Polygons

Using a drawing implement (e.g. graphite pencil, pen and ink, crayon or paint and brush) of your choice, and without the assistance of geometrical instruments (e.g. ruler, pair of compasses and set squares), draw the following polygons: a regular pentagon, a regular hexagon and a regular octagon.

2e   A Collection of Regular All-Over Patterns Using Polygons

Trace or scan each of the regular polygons produced in Exercise 2d. From these copies, using a balanced palette of up to five colours (plus black and white should you choose to use these), create a collection of three regular all-over patterns (one from pentagons, one from hexagons and one from octagons). In each case, the constituent polygons may be of various sizes, free standing or overlapping. In an accompanying single sheet, state a means of production, the anticipated end use (only select one end use for your collection) and the intended raw materials to be used. Remember to give the viewer an idea of the intended scale for your designs (so, a scale of 1:2 would indicate that the patterns actually produced would be twice the size of the paper copies).

2f   A Collection of Regular All-Over Patterns Using Circles

Hand draw (so no geometric instruments are permitted) three circles of various diameters. Trace or scan each of these, and using a balanced palette of up to five colours (plus black and white, should you choose), create a collection of five regular all-over patterns, using mixtures of the three circle sizes, overlapping or free-standing. Also, produce an accompanying mood/story board identifying the anticipated end use, the means of production and intended raw-materials to be used. Feel permitted to edit or resize your images to meet the requirements of manufacture or anticipated end use. Remember to indicate to the viewer the intended scale for your designs; this can be done on your mood/story board, using various available software to 'map' a copy of an image of one of your regular patterns onto an outline or photograph of the intended product.

REFERENCES

Ambrose, G. and P. Harris. 2008. *Basics. Design 07: Grids*. Lausanne (Switzerland): AVA Publishing.

Behrens, R. R. 1998. Art, design and gestalt theory. *Leonardo*, 31 (4): 299–303.

Betts, R. J. 1993. Structural innovation and structural design in renaissance architecture. *Journal of the Society of Architectural Historians*, 52 (1): 5–25.

Bidwell, J. K. 1986. A Babylonian geometrical algebra. *The College Mathematics Journal*, 17 (1): 22–31.

Birkett, S. and W. Jurgenson. 2001. Why didn't historical makers need drawings? Part I–practical geometry and proportion. *The Galpin Society Journal*, 54: 242–284.

Brunes, T. 1967. *The Secret of Ancient Geometry and Its Uses*. 2 vols. Copenhagen: Rhodos.

Chilton, J. 1999. *Space Grid Structures*. Oxford: Architectural Press.

Collins, P. 1962. The origins of graph paper as an influence on architectural design. *Journal of the Society of Architectural Historians*, 21 (4): 159–162.

Dabbour, L. M. 2012. Geometric proportions: The underlying structure of design process for Islamic geometric patterns. *Frontiers of Architectural Research* 1: 380-391.

Day, L. F. 1999 [1903]. *Pattern Design*. New York: Dover and, previously, London: Batsford.

Dudley, C. J. 2010. *Canterbury Cathedral: Aspects of Its Sacramental Geometry*. Bloomnington (IN): Xlibris.

Edwards, C. 2009. *How to Read Pattern*. London: Herbert Press.

Edwards, E. B. 1967 [1950]. *Pattern and Design with Dynamic Symmetry*. New York: Dover.

Elam, K. 2001. *Geometry of Design*. New York: Princeton Architectural Press.

Elam, K. 2004. *Grid Systems: Principles of Organizing type*. New York: Princeton Architectural Press.

Halmekoski, P. 2014. *Drawing Circle Images*. Espoo (Finland): Deltaspektri.

Hambidge, J. 1967 [1926]. *The Elements of Dynamic Symmetry*. New York: Dover Publications and previously New York: Brentano's Inc.

Hann, M. 2012. *Structure and Form in Design: Critical Ideas for Creative Practice*. Oxford: Berg.

Hann, M. and I. S. Moxon. 2019. *Patterns: Design and Composition*. New York: Routledge.

Hiscock, N. 2007. *The Symbol at Your Door: Number and Geometry in Religious Architecture of Greek and Latin Middle Ages*. Aldershot, UK: Ashgate.

Hurlburt, A. 1978. *The Grid: A Modular System for the Design and Production of Newspapers, Magazines, and Books*. New York: Van Nostrand Reinhold.

Jacobs, H. R. 1974. *Geometry*. New York: W. H. Freeman.

Jacobson, D. M. 1986. Hadrianic architecture and geometry. *American Journal of Archaeology*, 90 (1): 69–85.

Kandinsky, W. 1979 [1926]. *Point and Line to Plane*. New York: Dover. Previously as *Punkt und Linie zu Fläche*. Weimar: Bauhaus Publications.

Livio, M. 2002. *The Golden Ratio*. New York: Broadway Books.

Lundy, M. 2001 [1998]. *Sacred Geometry*. New York: Walker and Company.

Müller-Brockmann, J. J. 2008. *Grid Systems in Graphic Design: A Visual Communication Manual for Graphic Designers, Typographers and Three Dimensional Designers*. Sulgen/Zürich: Niggli.

Oei, L. and C. De Kegel. 2002. *The Elements of Design: Rediscovering Colours, Textures, Forms and Shapes*. London: Thames & Hudson.

Özdural, A. 2000. Mathematics and arts: Connections between theory and practice in the medieval Islamic world. *Historia Mathematica*, 27: 171–201.

Popovitch, S. 1924. Conception of space in old masters. *The Burlington Magazine for Connoisseurs*, 44 (254): 222+224–228.

Pough, A. 1976. *Polyhedra: A Visual Approach*. Los Angeles: University of California Press.

Poulin, R. 2011. *The Language of Graphic Design: An Illustrated Handbook for Understanding Fundamental Design Principles*. Gloucester (MA): Rockport Publishers.

Puhalla, D. M. 2011. *Design Elements: Form and Space*. Beverly (Massachusetts): Rockport Publishers.

Rich, B. 1963. *Plane Geometry*. London: McGraw-Hill.

Roberts, L. and J. Thrift. 2002. *The Designer and the Grid*. Brighton (UK): RotoVision.

Samara, T. 2003. *Making and Breaking the Grid: A Layout Design Workshop*. Gloucester (MA): Rockport Publishers.

Steed, J. and F. Stevenson. 2012. *Sourcing Ideas*. London: Bloomsbury.

Stewart, M. 2009. *Patterns of Eternity*. Edinburgh: Floris Books.

Thompson, M. 1977. Computer art: Pictures composed of binary elements on a square grid. *Leonardo*, 10 (4): 271–276.

Wang, C. 2017. The Geometric Division of Space. A Framework for Design Analysts. PhD thesis, University of Leeds (UK).

Watts, C. M. and D. J. Watts. 1987. Geometrical ordering of the garden houses at Ostia. *Journal of the Society of Architectural Historians*, 46 (3): 265–276.

Watts, D. J. and C. M. Watts. 1986. *A Roman Apartment Complex in the Second-Century Garden Houses of Ostia. Scientific American, Inc*, 132–139.

Watts, D. J. and C. M. Watts. 1992. The role of monuments in the geometrical ordering of the Roman master plan of Gerasa. *Journal of the Society of Architectural Historians*, 51 (3): 306–314.

Wiemer, W. and G. Wetzel. 1994. A report on data analysis of building geometry by computer. *Journal of the Society of Architectural Historians*, 53 (4): 448–460.

Wightman, G. 1997. The imperial fora of Rome: Some design considerations. *Journal of the Society of Architectural Historians*, 56 (1): 64–88.

Wong, W. 1972. *Principles of Two-Dimensional Design*. New York: Van Nostrand Reinhold.

CHAPTER 3

TYPES

3.1 INTRODUCTION

It was noted in Chapter 1 of this present book that regular patterns may be classified in various ways, including by reference to their thematic content. Justema (1976), for example, in his review of pattern varieties included headings such as animal, figurative, floral and geometric. While many of the pattern types illustrated in this present book may well fit conveniently under these or similar headings, the intention is different. Rather, in addition to differentiating between regular border patterns and regular all-over patterns, the focus will be on structural characteristics and how the component parts are arranged. Some important motif types, considered often to be the building blocks of regular patterns, are identified also. Regular patterns, such as spots, stripes and checks, largely ignored by analysts, will be given attention here also, as well as regular medallion and diaper patterns. Relevant literature is identified as well.

3.2 MOTIFS

Numerous handbooks or compendia illustrating motifs, logos, trademarks, symbols or single non-repeating images have been published over the years. Probably foremost among these was that produced by Petrie (1990 [1930]), entitled *Decorative Patterns of the Ancient World*, with the word 'pattern' used by Petrie to include 'all the decorative imaginings' (1990, p. 3) of human beings; this included both regular patterns as well as large quantities of motifs or stand-alone figures. Some later compendia of motif-like figures were concerned with aspects of visual identity, some with the design process and some with advertising, marketing, branding or related matters. Often motifs of one kind or another have been used as logos (associated with a company or institution) or devices (such as road signs, giving instructions or indicating the possible presence of certain phenomena). Some too, on their own or in combination with other motif types, have been used as the building blocks of regular border or regular all-over patterns.

A substantial, well-illustrated review of sign types, mainly sourced from twentieth-century urban environments, was provided by Ballinger and Ballinger (1972). A notable compendium was provided by Hornung (1946 [1932]). Valentine (1965), in a review of forms of embellishment in medieval manuscripts, provided a worthwhile glossary, which included many motif types. There are also numerous compendia of motifs featured on products of various kinds; in the textiles context, probably foremost among these are publications dealing with carpet motifs and their symbolism. Bennett (1977), Denny (1979) and Stone (1997) are typical. Examples of carpet motifs from various collections, cross-referenced with these publications are presented in Figures 3.1 to 3.20. Symbolism and interpretation varies from source to source and probably also from culture to culture, so no attempt is made at listing possible symbolism of the examples given. Often motifs such as these would form part of a repeating unit, either in regular border form or, occasionally, as part of a regular all-over pattern.

Common sense suggests that in the historical stages of development, visual motifs, signs or emblems evolved first. It is unknown, however, whether regular border patterns predated the use of regular all-over patterns, as the former may have been introduced to enhance the presentation of a larger-scale non-repeating unit of some kind or, alternatively, may have been introduced after regular all-over patterns and were used simply as a means of enhancing these.

(Text continued after figures)

Figure 3.1– Figure 3.20
There are numerous books dealing with carpets. Bennett (1977), Denny (1979) and Stone (1997) are typical examples. Examples of carpet motifs from various collections, cross-referenced with these publications are presented in Figures 3.1 to 3.20. Symbolism and interpretation varies from source to source and probably also from culture to culture, so no attempt is made at listing possible symbolism of the examples given. Often motifs such as these would form part of a repeating unit either in regular border form or, occasionally, as part of a regular all-over pattern.

Figure 3.1

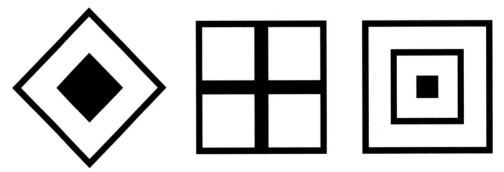

Figure 3.2

Figure 3.3

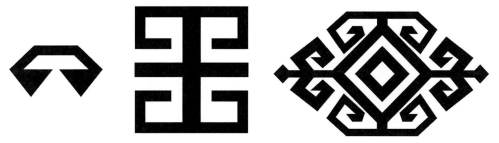

Figure 3.4

Figure 3.5

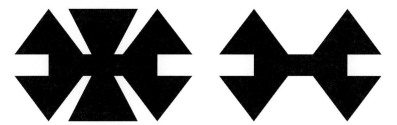

Figure 3.6

Figure 3.7

Figure 3.8

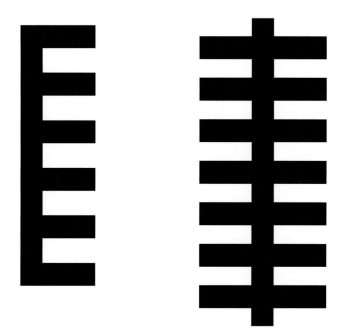

Figure 3.9

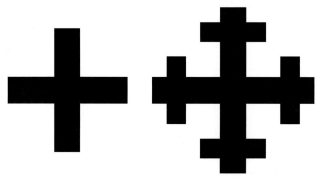

Figure 3.10

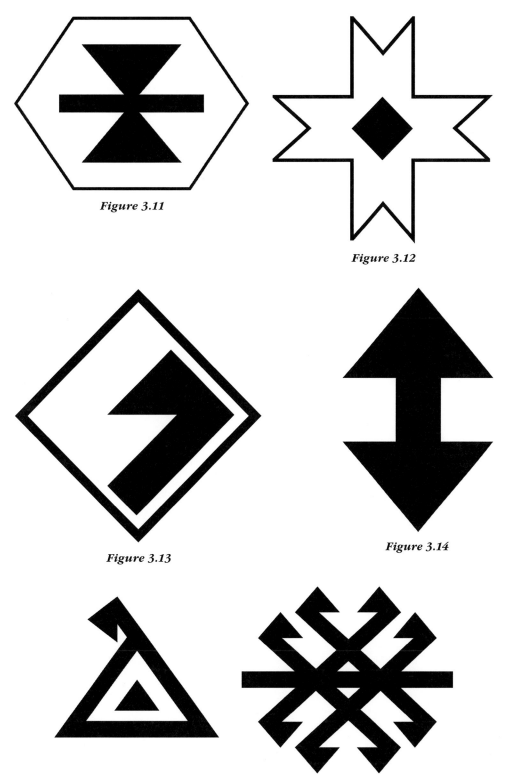

Figure 3.11

Figure 3.12

Figure 3.13

Figure 3.14

Figure 3.15

Figure 3.16

Figure 3.17

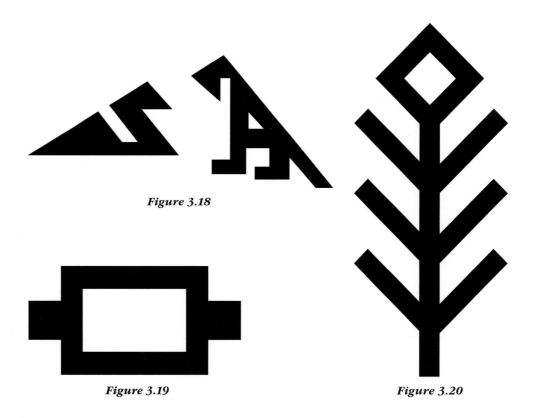

Figure 3.18

Figure 3.19

Figure 3.20

3.3 *REGULAR BORDER PATTERNS*

Astley (1990) provided a profusely illustrated book with 110 plates, many depicting two, three or four regular border patterns. Possibly the most extensive reference to regular border patterns was made by Fenn (1993 [1930]), an undoubted twentieth-century expert in their construction. He noted the value of regular border patterns throughout history and across cultures and their use in ceramics, metal work, embroidered and woven textiles, illuminated manuscripts, advertisements and in various architectural contexts, including plaster mouldings and wood carvings (Fenn 1993 [1930], pp. 18–21). Various characteristics were highlighted by Fenn (1993 [1930]) and a range of advice was given to practitioners. Probably most important was that repeating units should fit the space exactly and were not cut or divided, a condition which was relatively easy to achieve when a border was intended for the edge of a circular device such as a plate; in this latter case, it was simply a matter of segmenting the circle into six, eight or any other appropriate number and ensuring that the repeating unit was created to fit exactly the relevant segment. With rectangular shapes, the treatment of the corner is a challenge for the designer and, invariably, modification of the repeating unit is required. Fenn commented, however, that occasionally progress was straightforward: 'Many details based on the square will require little or no adaptation – the chequer or chess-board pattern is an example' (Fenn 1993 [1930], pp. 22–23). He recognised that the use of floral or plant-type motifs presented difficulties, as the 'growth should be consistent throughout' and should not reverse direction on the same branch nor should two stems be provided for a single flower (1993, pp. 24–29). As a rule, in all forms of regular-pattern design, it is best to ensure that plant growth projects upwards or laterally and only occasionally downwards. Fenn (1993 [1930]) presented several well-focused design solutions.

In regular border designs surrounding (or framing) a rectangle or square, the corner of the frame should be strong. Imagine an upward-placed rectangle with borders on each side. A common solution is to place identical units (such as a rosette) within a square at each corner. Meanwhile, as suggested by Fenn, the vertical side (or long) borders should be identical and column like and the horizontal upper and lower (or short) borders should be identical and 'frieze-like in character' (Fenn, 1993 [1930], p. 26). Importantly, Fenn (1993 [1930], p. 33) recognised that often borders required subsidiary borders, possibly consisting of a narrow band or one or more parallel lines; importantly too, he advised that the width of this subsidiary border should be in 'good proportion' and suggested that a good rule was one-sixth or one-eighth of the first border's total width (Fenn 1993 [1930], p. 33). Day (1999 [1903], p. 53) observed that a regular border 'may be described as confined always within fixed marginal (usually parallel) lines, which, whether expressed or understood, determine its depth or breadth'. One particular regular border-pattern type which attracted the attention of Day (1999 [1903], pp. 57–59) was the fret or key pattern 'found among Chinese and Mexicans, [and] among Greeks and Fiji islanders'. He considered that the degree of refinement achieved by the Greeks deserved further attention. While Day announced the 'futility' of simply copying the Greek fret, he nevertheless recognised that it had qualities of 'balance' and 'simplicity' of importance to regular pattern designers (Day 1999 [1903], p. 59).

3.4   STRIPES AND CHECKS

Lines were considered briefly in Chapter 2, and it was argued that these are the foundation for all shapes. They are considered to have direction and collections of lines are often, though not always, parallel. Regular stripe patterns have a directional and parallel aspect and consist of regularly repeating series of parallel lines of definite width and of definite distance apart. Regular stripe designs are widespread in many cultures and their development in some contexts may have occurred alongside early developments in weaving. Variations are numerous, with respect to width, order and orientation. Famously, vertical stripes suggest thinness and added height, whereas horizontal orientation implies added width. Treatises dealing specifically with stripes are surprisingly rare; most notable is O'Keeffe (2012). The present author (Hann 2015) presented a review of possibilities in the book *Stripes, Grids and Checks*. Stripe designs are of importance in weaving, and the issue is covered in most weaving text books. Regular-stripe effects are prevalent also in rolled-up resist-dyed textiles (where a cloth is rolled and bound tightly at preset distances around the roll, prior to immersion in a dye bath), as well as in knitting.

Pastoureau (1991) presented an interesting historical review of the applications of striped cloth, indicating its use as prison attire, as tailored suits (worn by city dwellers) and in the form of Pablo Picasso-like tee-shirts; he showed how, historically (in the Middle Ages, at least), stripes were associated with satanic worship but came to symbolise freedom and unity after the American and French revolutions.

Several constructions based on squares were identified in Chapter 2. Squares are of importance as components of regular design. When combined, in their simplest form such designs are known as 'checks', famously prevalent in Scotland in the form of Scottish clan tartans, a category of woven textiles. These are a surprisingly modern adaptation, with various regular check designs assigned to families or clans, based largely on the order of coloured yarns within the repeating unit (or sett). Relevant literature was identified in Hann (2015, pp. 78–82). Variations are produced in knitting, resist dyeing and printing. In the knitted variation, known as an 'Argyle check', the square component is turned through 45° and stretched.

A small selection of basic stripe and check designs is presented in Figures 3.21 to 3.25.

(Text continued after figures)

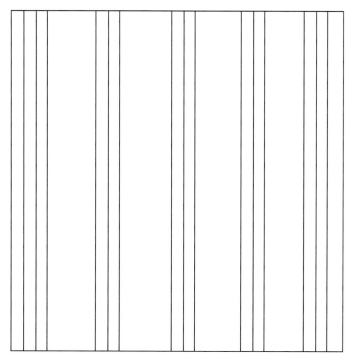

Figure 3.21
A basic stripe design.

Figure 3.22
A manipulated stripe design.

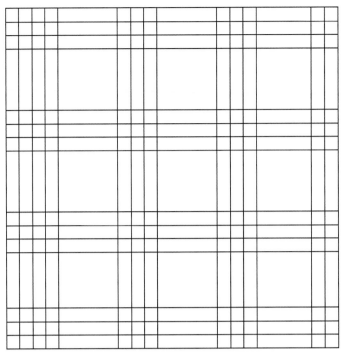

Figure 3.23
A basic check design.

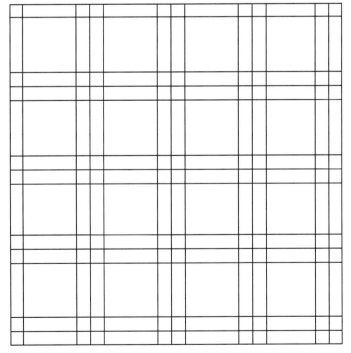

Figure 3.24
A basic check design.

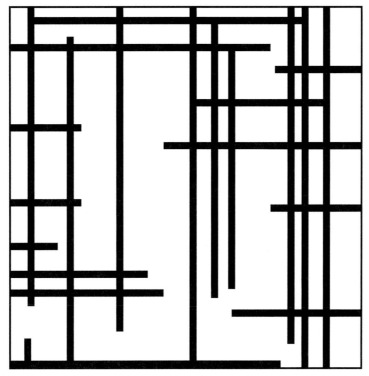

Figure 3.25
A manipulated check design.

3.5   SPOTS, MEDALLIONS AND DIAPERS

Treatises dealing exclusively with regular spot patterns are rare, though this design type is a common design category. The work by Hampshire and Stephenson (2006) is, however, useful, though they focused on circles and circular formations of various kinds and not specifically on regular spot patterns. Traditionally, in Japanese culture, the regular-pattern category known as *komon* included all 'small' regular patterns suited for printing on paper or textiles; many of these were of a spotted variety. In this present book, regular spot patterns will include all regular patterns with a circular, oval or similarly recognisable geometric figure, without obvious thematic content, invariably in flat colour against a contrasting background. Scale is not a consideration in this definition; so, in principle, the repeating unit of a regular spot pattern can range from very small (say, an arbitrary 1 centimetre) to very large (say, an arbitrary 1 metre). There will be obvious overlaps between regular spot patterns, as described here, and regular medallion patterns described further below, though it is suggested that common sense should prevail in placing a design under one heading or the other. It is recognised, however, that this may be a critical consideration among anthropologists or archaeologists when wishing to classify and compare one group of designs from one culture or time period with that from another, but from the viewpoint of designers, manufacturers and design analysts, the differentiation given should be sufficient. Various tie-and-dye techniques lend themselves readily to the production of regular spot patterns, and mechanised printing of the design type is probably in imitation of these previous craft achievements. Early technological developments in textile printing in Europe during the eighteenth and nineteenth centuries were stimulated by imports of printed textiles from Asia, mainly India, where tie-and-dye-type designs had been produced for centuries. It should be noted that occasionally fine spot patterns have been referred to as 'powdering' (Christie 1969 [1910], p. 65).

Regular medallion patterns are those patterns where the repeating component forms a clear foreground or island, consisting possibly of a known geometrical shape such as a circle, oval, square or regular hexagon, against a contrasting background. Unlike regular spot patterns, where the repeating component is of a single flat colour, regular medallion patterns can have embellishment, texture or a figured addition of some kind within the foreground (or medallion) component. It should be noted, however, as stated above, that such descriptions are arbitrary, and there will be overlap between the two regular-pattern types. 'Medallion' is a term given often to circular or oval shapes, and in the case of regular medallion patterns, the term is taken to mean a regular pattern consisting of equidistant circular, or diamond, or oval or any 'regular' polygon (such as a regular hexagon or regular heptagon) repeated systematically, without scale change, across the plane. Irregularly shaped convex polygons may be included also within the category; the important criterion is that a series of islands (or medallions) are presented against a contrasting sea (or background).

It is difficult to find a consensus on the precise nature of diaper patterns. Prior to its use in the United States to refer to a form of absorbent underwear, the word 'diaper' was used to refer to a pattern of repeated rhombic shapes and later came to describe a white cotton or linen fabric with this pattern. In the online Glossary of Medieval Art and Architecture, the term was used to refer to a pattern formed by small, repeated geometric motifs set adjacently to one another. A related definition is presented here: a regular pattern built on an equilateral triangle-, square- or diamond-shaped network. Diapers can be conveniently visualized as similar to tilings which tessellate (i.e. which cover the plane without gap or overlap). Although the word 'tessellation' is used generally when referring to tile arrangements, in the context of the present book, the term is applied also to denote repeating units of a regular pattern that are adjacent on all sides and, together, cover the plane without gap or overlap and thus have no background. Diaper patterns (all of which are considered here to have a tessellating feature) are thus different to regular medallion patterns (which show a motif against a contrasting background). In the context of this book, and when referring to regular all-over textile patterns, the word 'diaper' is used therefore to refer to regular patterns where the repeating units are located immediately adjacent to each other and share adjacent sides; so, a diaper can be considered as the 'regular' pattern equivalent of a regular tessellating tiling.

3.6 SUMMARY

This chapter explains briefly the nature of motifs and regular border patterns and provides further details on the characteristics of regular spotted, striped, checked, medallion and diaper patterns. In the chronology of aesthetic development, common sense suggests that spots and stripes may well have come at an early stage. Each could be seen in the natural environment, and each could come about through accidental mishap during a production process such as weaving or, maybe through a tie-and-dye-type procedure. Stripes may have emerged with the use of a potter's wheel where an impression can be left by one or more fixed points of some kind inserted initially in the clay together with a simple circular motion of the wheel itself, or during weaving with the ordering of yarns of different colour or count in the warp or weft direction. Initially, a certain random order may have been tried, but this would have given way soon to the human inclination towards regularity and order.

It can be speculated that familiarity with spots may have come around on observation of spatters of some kind, maybe during cooking. Spots and stripes would have been observed in nature and may well have been appreciated, but it seems unlikely that each came about deliberately through imitation but rather that, once each occurred during some form of human activity (maybe cooking, pottery, weaving or building), their worth might have been appreciated. So, although there was often the suggestion that spots and stripes were produced in imitation of nature, it is argued here that it is unlikely that this was the case. From spots, medallion forms of some kind may have developed with a thematic content or impression added subsequently, but maybe this followed on from pottery or some building craft. Stripes, if discovered during weaving, were probably simply straight and parallel, as is necessitated by conventional weaving. The development of checked textiles may have followed as an obvious further stage. Diamond-type renditions were probably a subsequent feature, maybe made obvious through knitting.

There could of course be further regular patterns which do not fit conveniently within the categories given above, though it is believed that the majority of regular patterns are indeed accommodated. A repeat of a regular all-over pattern can be discovered through connecting four 'equal' points. By 'equal' points is meant positions in the pattern which are oriented identically and are each surrounded by identical aspects. Four points connected will encase a full repeat of the regular all-over pattern.

EXERCISES

Dow (2007 [1920], p. 82) recognised that black against white provided strong contrast but, occasionally, appeared harsh. A remedy is to introduce a mid-way grey between the two extremes of black and white. Mix a mid-grey, using physical paint or digital mixing. You thus have three values: black, white and mid-grey. In years past, the alternative was to create an even wash of mid-grey on a stretched or pinned-down sheet of paper and, when dry, to paint with white and black on the washed surface; a similar approach was taken by the Italian Renaissance painters. In your responses to the exercises accompanying this chapter, you are required to use three values: black, white and mid-grey to 'colour' responses.

3a   A Collection of Motifs

From website or museum sources of your choice, collect twenty (photographed or hand-drawn) images of motifs which you believe could be of value in the development of regular patterns. Consider with care and select ten motifs. Present a line drawing of each of your selected ten and 'colour' using black and mid-grey against a white background.

3b   A Collection of Regular Border Patterns

From the ten motifs resulting from Exercise 3a, create a collection of six regular border patterns. Feel free to resize and edit as you wish. Identify an anticipated end use and means of manufacture.

3c A Collection of Regular All-Over Stripe Patterns

Consider stripe effects in the natural and manufactured/constructed environments. Collect up to fifty images (mainly photographs resultant from personal observations but supplemented, if necessary, by images from any other sources) which you feel express a visual stripe characteristic. Using a regular grid of your choice and using black, mid-grey and white only, create a collection of six regular all-over stripe patterns. Present each pattern on a single sheet of paper showing the intended size of your design. Also present a story/mood board showing a small selection of sources, and design-development images, the intended end use, anticipated means and cost of production, raw materials to be used and the anticipated price to the consumer.

3d A Collection of Regular All-Over Check Patterns

Consider check effects in the natural and manufactured/constructed environments. Collect up to fifty images (mainly photographs from personal observations but supplemented, if necessary, by images from any other sources). Employing a regular square grid and using black, mid-grey and white only, create a collection of six regular all-over check patterns, presented as specified in Exercise 3c and with an accompanying mood/story board as specified in Exercise 3c.

3e A Collection of Regular All-Over Spot Patterns

Consider spot effects in the natural and manufactured/constructed environments. Collect up to fifty images (mainly from personal observations but supplemented, if necessary, by images from any other sources). Employing a regular grid of your choice and using black, mid-grey and white only, create a collection of six regular all-over spot patterns, presented as specified in Exercise 3c and with an accompanying mood/story board, again as specified in Exercise 3c.

3f A Collection of Regular All-Over Medallion Patterns

Consider the nature of medallions (single non-repeating images) in the natural and manufactured/constructed environments. Using these observations, collect up to fifty images from any sources you may wish. Employing a regular grid of your choice and using black, mid-grey and white only, create a collection of six regular all-over medallion patterns. Present each regular pattern as specified in Exercise 3c. Also present a story/mood board to the specifications given in Exercise 3c.

3g A Collection of Regular All-Over Diaper Patterns

A regular all-over diaper pattern is considered to be a regularly repeating pattern with a repeating unit which shares sides/boundaries with all adjacent repeating units. Consider diaper-type effects in the natural and manufactured/constructed environments, and collect up to fifty images from any sources. Employing a regular grid of your choice and using black, mid-grey and white only, create a collection of six regular all-over diaper patterns. Present each pattern as specified in Exercise 3c together with a story/mood board, again as specified in Exercise 3c.

REFERENCES

Astley, S. 1990. *Border Designs*. London: Studio Editions.

Ballinger, L. B. and R. A. Ballinger. 1972. *Sign, Symbol and Form*. New York: Van Nostrand Reinhold.

Bennett, I. 1977. *Rugs and Carpets of the World*. London: Thames and Hudson.

Christie, A. H. 1969 [1910]. *Pattern Design. An Introduction to the Study of Formal Ornament*. New York: Dover and, previously, as *Traditional Methods of Pattern Designing*. Oxford: Clarendon Press.

Day, L. F. 1999 [1903]. *Pattern Design*. New York: Dover, and previously London: Batsford.

Denny, W. B. 1979. *Oriental Rugs*. Washington DC: Smithsonian Institute.

Dow, A. W. 2007 [1920]. *Composition. Understanding Line, Notan and Color*. New York: Dover and, previously, under title *Composition: A Series of Exercises in Art Structure for the Use of Students and Teachers*, New York: Doubleday Page and Company.

Fenn, A. 1993 [1930]. *Abstract Design and How to Create It*. New York: Dover and, previously, under title *Abstract Design: A Practical Manual on the Making of Patterns for the Use of Students, Teachers, Designers and Craftsmen*. London: Batsford.

Hampshire, M. and K. Stephenson. 2006. *Circles and Dots*. Mies (Switzerland): RotoVision.

Hann, M. 2015. *Stripes, Grids and Checks*, London and New York: Bloomsbury.

Hornung, C. P. 1946 [1932]. *Handbook of Designs and Devices*. New York: Dover and previously New York: Harper and Brothers.

Justema, W. 1976. *Pattern. A Historical Panorama*. Boston: New York Graphics Society.

O'Keeffe, L. 2012. *Stripes: Design Between the Lines*. London: Thames and Hudson.

Pastoureau, M. (translator: Jody Gladding). 1991. *The Devil's Cloth. A History of Stripes*. New York: Washington Square Press.

Petrie, F. 1990 [1930]. *Decorative Patterns of the Ancient World*, London: Studio Editions, and London: British School of Archaeology in Egypt, University College, and Bernard Quaritch.

Stone, P. F. 1997. *The Oriental Rug Lexicon*. London: Thames and Huidson.

Valentine, L. N. 1965. *Ornament in Medieval Manuscripts*. London: Faber and Faber.

CHAPTER 4

PARTITIONS

4.1 INTRODUCTION

The use of grids consisting of regular polygons appears to be long-standing historically. This chapter highlights the nature of the most common compositional frameworks used to assist pattern formation: particularly the regular grids based on the three regular tiling arrangements. It is well known that only three regular polygons can cover the plane on their own without gap or overlap: squares of equal size, equilateral triangles of equal size and regular hexagons of equal size. Each of the three is dealt with in this chapter. These three have traditionally formed the basis of grid forms used in the visual arts and design and especially in the design of regular patterns. The reason why there are only three possibilities is that only these three regular polygons when brought together (on their own) can create a vertex of 360° (4 × 90°; 6 × 60° and 3 × 120°), and so they can lie side to side without gap or overlap. These three possibilities, when presented in the form of tiling designs, are known as the 'regular' or 'Platonic' tilings. Further tiling arrangements, suited as compositional grids, and involving mixtures of two or more regular polygons, can be useful also as compositional grids in design, and these will be introduced later, in Chapter 6.

Grids are considered by many visual-arts-and-design practitioners to be fundamental to their activities. Although different reasons are given on why this is the case, it seems that the majority believe that grids give order to their work. A grid helps with the organisation and placement of components and their possible alignment. Using grids allows the arrangement of parts to be neat, clean and efficient and, if required, a hierarchy can be followed or given. A composition in sympathy with its parts can be achieved, rather than a random positioning of components. A cluttered layout of parts can be avoided.

Also in common use (mainly by those involved in web- or magazine-page design) is a grid of another type consisting simply of two parallel lines running at 90° to another set of two parallel lines, placed over a square or rectangle; the intersections created are regarded as of importance in the placement of components in any square or rectangular composition. This three-by-three arrangement with a total of nine divisions is explained further in Chapter 6, under the heading 'Single-Page Grids'.

4.2 SQUARES

Square formats dominate the world of the visual arts and design, from frames and outlines to constituent elements. Two series of parallel lines, one overlapping the other at 90°, can provide the most common framework used in the visual arts and design since ancient times. So, with simple square unit cells, an extensive grid across the plane can be produced. There is evidence that the square grid was in common use in ancient Egypt as a guideline to the proportions of figures depicted. It was used also during Italian Renaissance times for similar reasons and, in more recent times, has been used by regular-pattern designers as a guideline to ensure precise registration of components, in both printing and weaving.

It is common knowledge that woven textiles consist of two assemblies of parallel threads, one interlacing with the other at 90° (thus producing a grid of types based on a square). When each of the two sets of threads is arranged in a particular sequence of colours and then interlaced, a check fabric is produced. Although this procedure seems common worldwide, as noted previously (in Chapter 3), it was in Scotland where such fabric (predominately in wool) became known as a 'tartan' or 'plaid'; these textiles were associated closely with different family groups or regions depending on the order of colours (or sett). The various setts displayed by a collection of tartans held at the University of Leeds (UK), and in particular the ratios and proportions expressed in the collection, were examined by Hann and Wang (2016); based on these observations, a series of templates was proposed for use by designers. They found that the dominant ratios included 1:1, 1:2, 2:3, 1:3, 1:4 and 3:4, and a range of tiling designs conforming to these ratios was presented; these were then used as the basis for a collection of regular all-over pattern designs. Relevant illustrative material is presented in Figures 4.1 to 4.14.

(Text continued after figures)

Figure 4.1–Figure 4.14
Consideration by Hann and Wang (2016) of the collection of tartans held at the University of Leeds (UK) led to the development of a collection of regular all-over patterns, some depicted here together with underlying structures.

Figure 4.1

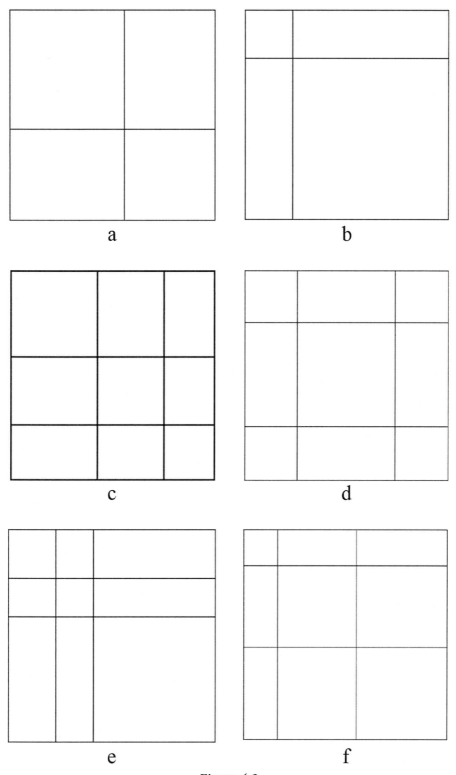

Figure 4.2

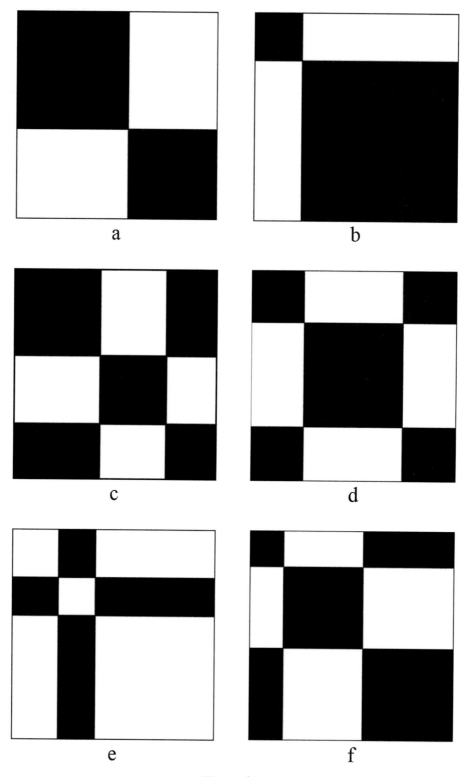

Figure 4.3

Figure 4.4

Figure 4.5

Figure 4.6

Figure 4.7

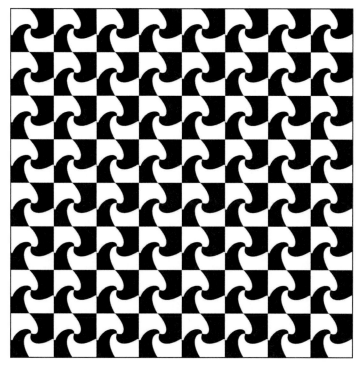

Figure 4.8

Figure 4.9

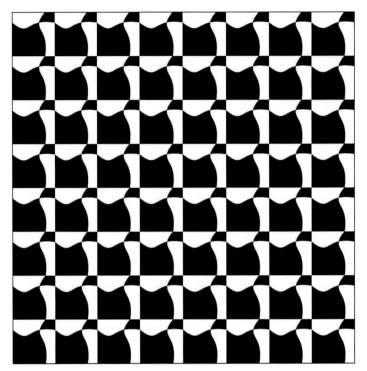

Figure 4.10

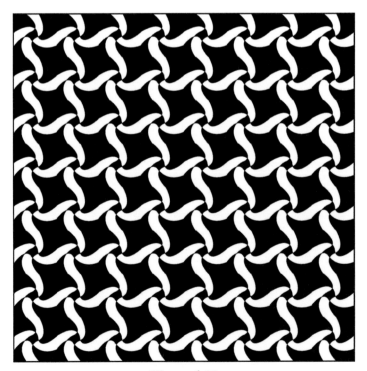

Figure 4.11

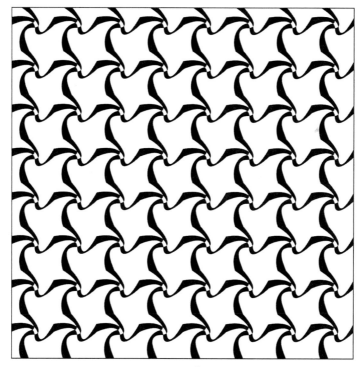

Figure 4.12

Figure 4.13

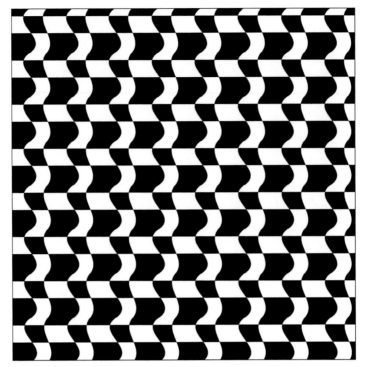

Figure 4.14

It is interesting that an original origami construction begins with a single piece of square paper, which is then folded (with no cutting or addition of paste permitted). When completed, origami pieces form a three-dimensional object which, when unfolded, yields an isosceles right-angled triangle (reported by Kappraff 1991, pp. 198–199) amongst other polygons. Such sources can be a fount of inspiration for visual-arts-and-design practitioners.

4.3   EQUILATERAL TRIANGLES

A series of three sets of parallel lines, with each set oriented at 60° to each other can create an equilateral-triangle grid. An alternative is to take equilateral triangles of equal size and cover the two-dimensional plane without gap or overlap. This structure has been used commonly as guidelines by visual artists and designers, particularly in tiling-design formats, but here its value as a compositional framework is acknowledged.

4.4   REGULAR HEXAGONS

Six equal-size equilateral triangles brought together, angle to angle and side to side, can produce a regular hexagon and, when brought into grid format with each unit cell consisting of one regular hexagon, a compositional grid of value in the planning and composition of regular patterns can be produced. As noted in the introduction to this chapter, equal-size-regular hexagons, like equal-size squares or equal-size equilateral triangles, can on their own tile the plane without gap or overlap.

4.5   PARTITIONS OF THE PLANE

The possibilities for subdividing the plane are limitless, and each result can offer alternative compositional frameworks for regular-pattern design and composition. It is well known that tiling the plane reached a sophisticated level of achievement in Islamic cultures (see, e.g. Broug 2008, 2013, 2016, for a masterly and sophisticated discussion of Islamic tiling arrangements). Star polygons are often a component part. The construction of these was reviewed by Kappraff (1991, pp. 169–173), showing that the extended sides of a pentagon intersect to create a star-shaped figure known as a pentagram (Kappraff 1991, p. 169). Star polygons can be produced using a circle as the structural format for each (Kappraff 1991, pp. 171–172).

Notation for tilings based on regular polygons was explained by Kappraff (1991, p. 173). Across each tiling arrangement, the vertex is identical. Where six equilateral triangles come together at each vertex, the relevant notation is {3,6}; where four squares come together the notation used is {4,4}; where three regular hexagons come together the notation is {6,3}. As noted previously in this chapter, each of the three formats has been used commonly as a compositional grid to arrange regular patterns. It is not possible to cover the plane without gap or overlap using equal-size regular pentagons, for example, though the great scientist Kepler obtained some interesting pentagonal arrangements. Kappraff (1991, p. 175) observed that it is possible to cover the plane using copies of any triangle or copies of any four-sided polygon. He observed also that any hexagon with opposite sides parallel and equal in length can cover the plane without gap or overlap (Kappraff 1991, p. 175).

Each of the three regular combinations using equilateral triangles or squares or regular hexagons has another associated tiling, known as a dual, created by placing a point at the centre of each unit cell and connecting these points with straight lines. An equilateral-triangle arrangement produces a regular hexagonal combination and a square arrangement produces another square arrangement. Thus, the duality of a {3,6} arrangement is {6,3}, {4,4} results in {4,4} and {6,3} results in {3, 6}. Further partitions of the plane using combinations of regular polygons are possible also. In tiling format, these are called the 'semi-regular' or 'Archimedean' tilings, and there are eight types in total. Although combinations rely on more than one polygon type, in each case the vertex is equal. Kappraff explained the basis of these constructions (Kappraff 1991, pp. 177–179). Tiling arrangements such as these have been used extensively historically, and Kappraff argued that their appeal was due to their inherent symmetry arrangements (Kappraff 1991, p. 178). Probably the most comprehensive treatise dealing with tiling arrangements was produced by Grünbaum and Shephard (2016 [1987]). Meanwhile the most extensive collection of tiling images was produced by Nicolai (2008). Kappraff showed several methods by which tiling arrangements could be transformed, through 'vertex motion', 'distortions', 'augmentation and deletion' and 'one-dimensional parquet deformation' (Kappraff 1991, p. 184). These and other forms of adjustment or manipulation are worth considering by pattern designers. The present author (Hann 2015, pp. 127–130) suggested that various forms of 'manipulation' of standard tiling arrangements were possible using readily available computer software; this is well worth considering, as common sense suggests that such tiling arrangements have been well resolved visually with unsuited arrangements having been removed from the catalogue a long time ago. So, the visual artist or designer is thus offered a ready-made framework on which to place components, and further manipulation of this framework using readily available software can offer yet another layer of originality.

Arrangements considered throughout much of this book repeat on a regular basis; that is, they are 'periodic'. There are, however, some arrangements which succeed in covering the plane yet do not repeat on a regular basis, and these are known as 'aperiodic'. It has already been noted that regular pentagons are unable to cover the plane on their own. However, it is worth noting that there is a class of tilings known as the 'aperiodic tilings' and foremost among these is a variety associated with Roger Penrose, the British mathematician, who proposed that the plane could be covered using only two types of polygon. Penrose dissected a regular pentagon and reassembled it into two polygons known as the 'kite' and the 'dart'. By means of precise rules of assembly, copies of the two polygons produce an overall pattern, without gap or overlay, which does not show regular repetition. However local points of five-fold rotational symmetry are apparent (though this does not extend in any regular way across the pattern). The result is shown in Figure 4.15. Another form of Penrose tiling uses two rhombuses to produce a similar aperiodic arrangement of the type shown in Figure 4.16. Again, various regular features may be detected, but these do not repeat following any apparent rule. A variety of further commentary on the nature of the composition was provided by Kappraff (1991, pp. 245–248). Again, copies of the two rhombuses are brought together in a very specific way (involving alignment of arrows and angles).

So, the arrangement proposed by Penrose is aperiodic (and does not show regular repetition).

Figure 4.15
A Penrose tiling (drawing by Chaoran Wang).

Figure 4.16
Another variety of Penrose tiling (drawing by Chaoran Wang).

4.6   SUMMARY

This chapter discussed the nature of the unit cells associated with the three common regular grid types used on their own in regular all-over pattern construction: equal-size squares, equal-size equilateral triangles and equal-size regular hexagons. Various further partitions of the plane were identified and discussed briefly, including Penrose-type arrangements, created from only two tiles and arranged specifically to be aperiodic (or non-repeating).

EXERCISES

4a   Regular All-Over Patterns from a Square Grid

Using a grid of equal-size square unit cells, and employing the ten motifs selected from those collected as part of the response to Exercise 3a, produce a collection of five regular all-over patterns for an end use of your choice (which must be specified within your submission), using a means of production of your choice and employing only three colour values (black, white and mid-grey). Feel free to edit, resize, extend or overlap motifs as you may wish.

4b   Regular All-Over Patterns from an Equilateral-Triangle Grid

Using a grid consisting of equal-size unit cells of equilateral triangles, and using the ten motifs selected as part of your response to Exercise 3a, create a collection of five regular all-over patterns following the specification given for Exercise 4a.

4c Regular All-Over Patterns from a Hexagonal Grid

Using a grid consisting of equal-size unit cells of regular hexagons, create a collection of five regular all-over patterns following the specifications given for Exercise 4a.

4d In the Tradition of Mondrian

Review and consider carefully the paintings of the twentieth-century artist Piet Mondrian, particularly his compositions consisting of square or rectangular components coloured often using white, black and the three primary colours (blue, red and yellow). The specifications of this exercise are different, but you could benefit from careful consideration of Mondrian's work. Ask these and similar questions: Why did the artist divide his canvas into rectangles? Did the artist follow any rules of proportion? Why did the artist often use black and white and three primary colours (though there were exceptions to this)? Why was Mondrian's work considered to be attractive? What was/ is the appeal? With these considerations in mind, you are required to select one of the three regular grids (consisting of unit cells of equal-size squares, equal-size equilateral triangles or equal-size regular hexagons) and to divide each unit cell identically with up to six straight lines and to 'colour' each (identically) using a palette of up to five colours, with the proviso that adjacent areas (with a common side or part of a side) are coloured differently (though you are permitted to 'outline' each division with black lines, should you wish).

4e Partition of Equilateral Triangle

Hand draw an equilateral triangle (to side length of your choice). Divide the figure by connecting (hand-drawn) straight lines from each angle to the midway point on the opposite side, thus creating a figure with six component parts. 'Colour' each part with a single colour from a palette of up to five colours (plus black and white should you choose). Rotate (and repeat) the 'coloured' triangular unit, with one of its vertices as centre, six times, till it coincides with itself. You have thus created a hexagon from six triangular units. Take this six-part 'motif' and repeat across the plane to create a regular all-over pattern.

4f Partition of Square

Hand draw a square (to side length of your choice), and divide with two (hand-drawn) straight diagonals from each of two angles to its opposite angle and two (hand-drawn) straight lines from the midway point on each of two sides to the midway point on its opposite side, thus creating a figure with eight component parts. Follow procedures outlined for Exercise 4e, but in this case, rotate the 'coloured' figure four times to create a larger square.

4g Partition of a Rectangle

Hand draw a rectangle of proportions 2:1 (thus consisting of two squares). Connect opposite angles and midway points on opposite sides with (hand-drawn) straight lines to create a figure with eight interior divisions. 'Colour' each division with a selection from a palette of up to five colours (plus black and white should you choose), ensuring that adjacent areas are 'coloured' differently. Place your 'coloured' rectangle in a grid format (in block, half-drop or brick orientation) and repeat across the plane to create a regular all-over pattern.

REFERENCES

Broug, E. 2008. *Islamic Geometric Patterns*. London: Thames and Hudson.

Broug, E. 2013. *Islamic Geometric Design*. London: Thames and Hudson.

Broug, E. 2016. *Islamic Design Workbook*. London: Thames and Hudson.

Grünbaum, B. and G. C. Shephard. 2016 [1987]. *Tilings and Patterns*. New York: W. H. Freeman.

Hann, M. 2015. *Stripes, Grids and Checks*. London: Bloomsbury.

Hann, M. A. and C. Wang. 2016. Symmetry, ratio and proportion in Scottish clan tartans. *The Research Journal of Costume Culture*, 24 (6): 873–885.

Nicolai, C. 2008. *Grid Index*. Berlin: Gestalten.

Kappraff, J. 1991. *Connections: The Geometric Bridge between Art and Science*. New York: McGraw-Hill.

CHAPTER 5

SYMMETRY

5.1   INTRODUCTION

It is recognised commonly that symmetry is a dominant structural feature found throughout the natural, built and manufactured environments. Embellishing objects with symmetrical images has been practised for thousands of years, but precisely when the first symmetrical images were produced remains unrecorded and is still debated. A concise, though well-focused, review was provided by Hahn (1998). In the late-twentieth century, a two-volume compendium, edited by Hargittai (1986, 1989), contained more than 100 papers, each focused on one aspect of the subject and, in the early twenty-first century, an extensive bibliography, taken from a surprisingly wide range of subject disciplines, was published by Washburn and Crowe (2004). So, the phenomenon manifests itself in a wide range of contexts. In the popular imagination, symmetry is considered to be present when an object, image or motif consists of two component parts, one a reflection of the other. While the two parts are of equal weight, the contents and shape may be oriented in opposite directions; still, in the bulk of literature, each reflected component is regarded as 'equal' (and, occasionally, it is indeed the case that component parts of a symmetrical whole are actually indistinguishable visually, as would be the case with a reflection axis through the midway points at opposite sides of a square).

So, in popular understanding, the word 'symmetry' is used to refer specifically to reflection symmetry, particularly where all parts on the left-hand side of an imaginary (two-sided) mirror are reflections to all parts on the right-hand side. This is known as 'reflection symmetry', with one reflection only. However, higher orders of reflection symmetry are possible, using two, three or more reflection axes. Visualisation is helped if each object, motif or figure is imagined as inscribed within a circle; each successive reflection axis reflects from two sides and is imagined as passing through the centre of that circle. So, in the popular imagination, at least, shapes or objects may be described as 'symmetrical' if they can be split into two or more reflected parts.

Rotational characteristics are another possibility; this is where each successive part is a rotation of the part before. Again, it is helpful to imagine the object, image or motif inscribed within a circle. Two-fold rotational symmetry is when the component part rotates through 180° and then rotates through a further 180° to coincide with its original self; three-fold rotational symmetry is when rotation is through 120°, a further 120° and then a final 120° to coincide with its original self. A clear division of 360° is therefore necessary. So, four-fold rotational symmetry involves four rotations, each of 90°; six-fold rotational symmetry involves six rotations, each of 60°. Of course, it should be recognised that an object, image or motif may instead be considered asymmetrical, in which case each consists of a single component part only, and reflection or rotational characteristics cannot be detected.

Hann and Thomson (1992) reviewed the development of symmetry concepts and perspectives and gave numerous examples of symmetrical motifs, regular border and regular all-over patterns (each classified with respect to its symmetry characteristics) and explained also the nature of two-colour counter-change motifs and patterns, showing how these too may be classified with respect to their inherent symmetry characteristics. Where systematic repetition is a component characteristic in motifs and patterns, then symmetry classification is a possibility. It has been shown by scholars such as Washburn and Crowe (1988) how symmetry classification can be used to classify motifs and patterns from different cultural or historical contexts and, by the early twenty-first century, it appeared to be accepted widely that different cultures expressed their own unique symmetry preferences and that, assuming the availability of representative data, symmetry classification was a valid means of detecting change and variation in the design preferences of different cultures and/or historical periods. Despite this apparent widespread acceptance, however, the use of symmetry classification as a means of handling cultural, historical and/or archaeological data has not been as widely used as a means of data analysis as would be expected; this may well be due to the belief (mistaken, in the view of the present author) that a means of classification by symmetry characteristics is too complex for widespread use. For this reason, it seems appropriate to trim the explanation of symmetry concepts to the absolute minimum; this is done below. Brief reviews of symmetry concepts and principles, together with relevant illustrative material, were provided previously by Hann and Thomson (1992), Hann and Thomas (2010) and Hann (2012, pp. 72–96).

Significant contributions have been made by anthropologists to understanding the nature of the visual arts and design in different cultural contexts. The geometric principles of symmetry have been used to bring enlightenment to the understanding of motifs and regular patterns in textiles, ceramics, basketry and other forms of crafted product. An explanation of the development of this and related perspectives can be seen in Washburn (1983). A later work, edited by Washburn (2004), recognised the importance of symmetry to aspects of cultural identity. Mention should be made also of Washburn and Crowe (the former an anthropologist and the latter a mathematician) who, in their 1988 seminal text *Symmetries of Culture*, explained the nature of geometric symmetry and how it may be used in the classification of motifs and regular patterns from different cultural or historical contexts; they showed that the analysis of data from different sources expressed different symmetry preferences. Of great further importance was the realisation that symmetry classification was a culturally sensitive tool (shown through non-random distributions of symmetry preferences from a finite number of possibilities for each group of data examined). Further to this, they argued that, as symmetry classification was a highly objective means of characterising the regular patterns of different cultures, it followed that different researchers could independently obtain similar results from the same set of data, a possibility that was remote previously using subjective appraisal.

The principle of symmetry is of universal significance and has relevance across the visual arts, design, architecture, mathematics, engineering and physical and biological sciences. From the viewpoint of the visual arts and design, notions of periodicity, regularity and repetition are of importance and symmetry-related concepts can be used to enhance our understanding of these.

This chapter identifies some of the main literature concerned with symmetry and, in particular, with symmetry in motifs and regular patterns; it also highlights how such knowledge when used in classification can help to order data and be used to compare data from different cultural or historical sources, a research procedure reproducible from one researcher to another.

5.2   CONCEPTS, ORIGINS AND DEVELOPMENTS

Pattern designers have, in general, been aware of the crucial role played by measurement and geometry in the organisation of regular patterns. Thus, over the course of the late-twentieth century, there were monumental strides in the understanding of symmetry, including the symmetry aspects of patterns. The bulk of explanatory literature has not been intelligible, however, to the majority of pattern designers, due principally to the inaccessibility of mathematically oriented texts and the use of unfamiliar symbols and terminology. Advances in mathematical symmetry have indeed contributed much to the classification, differentiation and understanding of patterns and their construction; scholars such as Walker and Padwick (1977), Stevens (1981), Schattschneider (1978) and Washburn and Crowe (1988) have done much in their attempts to alleviate the challenge faced by many designers, by introducing methodical and readily understandable approaches for the non-mathematically aware. From the viewpoint of the mathematically unaware, probably the most readily accessible and understandable of explanatory books dealing with pattern symmetry was Jackson's *How to Make Repeat Patterns*, published in 2018.

Schauermann's (1892) publication is a forgotten gem from the viewpoint of those with an interest in understanding the nature of regular patterns. A vast range of issues of relevance to regular tilings and patterns was dealt with, and a recognition was made of various symmetry combinations long before their announcement by scientists in the next century. It seems that Schauermann (1892) managed to capture the majority of concepts of value to considering symmetry in patterns. However, his terminology is obscure and his illustrative material confusing.

Subsequently, for much of the twentieth century, perspectives of relevance to the understanding of patterns came from physicists and crystallographers who were productive in elucidating and classifying the underlying structure of naturally occurring substances, especially crystals, and it is from these origins that symmetry considerations were adopted and used, initially, in the consideration of archaeological material. Relevant literature was identified by Washburn and Crowe (1988).

Mention should be made also of Woods (1935a, b, c, 1936) who, although a mathematician, succeeded in using crystallographic concepts to classify motifs and regular border and all-over patterns; working in the Textile Department of the University of Leeds (UK), he focused on presenting a rational system of classification that could be used by textile designers and others not concerned directly with the scientific aspects of textile manufacture. From these sources, it was established that all symmetrical motifs, as well as regular border and all-over patterns, could be classified by reference to their symmetry characteristics.

Schattschneider (1978) provided a fully comprehensive review of symmetry concepts and principles; this article is recommended highly to readers wishing to develop knowledge of symmetry in regular patterns. Swenson (1978) provided early guidelines on how to proceed with examining the structures underpinning ancient artefacts (in this particular case, manuscript pages). Stevens (1981) offered a fairly comprehensive introduction to symmetry concepts, illustrated by an extensive, worldwide collection of regular patterns. Locher (1982) provided a biography of M. C. Escher with illustrations of his graphic work, including many well-known tessellating designs. At a later date, Schattschneider's (1990) *M. C. Escher. Visions of Symmetry* stands as the most comprehensive (and inspirational) twentieth-century publication dealing with Escher's work.

A characteristic thoroughness was provided by Grünbaum and Shephard (1992) in their consideration of interlacement designs from Islamic sources; theirs is a publication worth making reference to in the quest to understand the symmetry characteristics of regular patterns more fully. In this regard, it is relevant also to refer to the earlier work of Hankin (1934), who did much to develop an understanding of the geometric characteristics of regular patterns in general and, in particular, possible construction techniques used. McDowell (1994) explained symmetry and associated concepts using a large number of quilt illustrations, and Yang (1996) presented a comprehensive review of the various symmetry-related developments, with the focus mainly on how these may have influenced thinking in the area of physics.

The word 'tessellations' is often found in discussions of symmetry. These are distributions of tiles which fit without gap or overlap. Fathauer (2008) offered a masterly discussion of tessellations and step-by-step explanations with illustrations on how to create regular tessellating designs. Each chapter in his treatise listed activities for use by teachers in a classroom setting, giving numerous original examples of tessellations, with many based on square grids, including tessellations with translational, rotational and glide-reflectional symmetries; a closely related grid of rectangles (made possible by simply stretching the design in one direction) was introduced also. A series of templates to assist the reader in creating original regular all-over tessellating patterns was included, and grids formed from equilateral triangles or hexagons were considered as well.

Knight (1998) presented a good, readily understandable explanation of symmetry concepts and classification, highlighting also some of the apparent ambiguities in conventional symmetry classification. Endress (1999) presented a comprehensive review of the presence of symmetry in flowers and the development of relevant observations among scholars. Conway et al. (2008) provided an extensively illustrated book which included explanations of the mathematical aspects of symmetry in a way largely understandable to the non-specialist, and Farris (2015) provided a richly illustrated explanatory text, ideal for students and practitioners with some knowledge of mathematics; for those without such knowledge, the illustrative material in the latter publication is certainly attractive and engaging and is well worth consultation. Broug (2008, 2013, 2016) provided a masterly analysis and discussion of the nature of tilings and patterns from Islamic sources. Having reviewed briefly some of the key literature, further attention is directed below to a range of relevant concepts.

As noted previously in the introduction to this chapter, various well-known geometrical figures have an obvious reflection symmetry, including an equilateral triangle, with three (imaginary) reflection axes going through its centre, a square with four, a regular pentagon with five, a regular hexagon with six, etc. A further type of symmetry (again, as noted in the introduction to this chapter), where the rotational properties of objects are recognised, is rotational symmetry, which identifies the number of identical components of an object which would coincide exactly with each other if rotated by a given number of degrees. So, if a motif permits rotation of half of its content through 180° and all the components of that half coincide exactly with the other half, then the motif is deemed to have two-fold rotational symmetry. So, as mentioned previously in the introduction to this chapter, rotational symmetry is best imagined within a circle, and where rotation needs to be three times within that circle for all components to coincide and for the motif to return to its starting point, then the motif is deemed to have three-fold rotational symmetry. It was noted previously, also, that higher orders of rotational symmetry are possible, and there are no restrictions on orders of rotation and reflection when considering motifs or other two-dimensional figures.

With regular border patterns and regular all-over patterns, there are various geometrical realities which impose restrictions. Prior to further mention of these it should be stated that both regular border and all-over patterns can exhibit two further symmetry characteristics: translation and glide-reflection. So, it should be noted at this stage that the four basic symmetry operations of reflection, rotation, translation and glide-reflection can be brought together in seven distinct ways to provide the structural basis for all regular border patterns and in seventeen distinct ways to provide the structural basis for all regular all-over pattern types. Schematic illustrations of the four symmetry operations are given in Figure 5.1. Jackson considered reflection symmetry to be present when 'one half of a figure' was 'the mirror image of the other half' (Jackson 2018, p. 30); rotation to be the 'duplication' of a figure 'around a central point' (Jackson 2018, p. 22); translation to be 'the movement of a figure' by a specified distance in a particular direction, noting also that the 'distance and angle between the figures' remained constant and the 'figures neither rotate or reflect' (Jackson 2018, p. 26); glide-reflection to be the combination of two symmetry operations, with translation symmetry 'followed by a reflection symmetry operation in the same direction as the translation' (Jackson 2018, p. 34).

Translation is therefore a simple repetition whereby a motif is 'translated' from a specific location (without change in size, orientation or content) to another position (considered in one direction only with regular border patterns and in two distinct directions with regular all-over patterns). Meanwhile, glide-reflection is an action which combines translation with reflection, best imagined as the imprint made by feet walking on wet sand. With border patterns, assuming an orientation in a horizontal direction, reflections in two directions only (horizontal and vertical) are possible, glide-reflection in a horizontal direction only is possible and rotation can only be two-fold. With regular all-over patterns, the principal restriction is that rotation can only be two-fold, three-fold, four-fold and six-fold. As noted above, combinations of these four symmetry operations permit seven regular border pattern types and seventeen regular all-over pattern types.

Symmetry classifications of representative groups of patterns have been made by several researchers, though the pioneers of symmetry analysis in cultural contexts are undoubtedly Washburn and Crowe (1988, 2004). As noted previously, when representative groups of patterns from different cultural or historical contexts are analysed with respect to their symmetry characteristics, different symmetry preferences are expressed by each distinct group.

5.3   SYMMETRY IN MOTIFS

With rotation symmetry, the equal component parts are presented at regular intervals around a central rotational point. With reflection symmetry, equal components are reversed across a central reflection axis or mirror line. In its most basic form, rotational symmetry involves only two equal components around a rotational centre.

As noted previously in the introduction to this chapter, in the popular imagination, objects or images with two parts, one a reflection of the other, are regarded as 'symmetrical'. It has been noted also that, when considering motifs and patterns, the concept of symmetry can be extended beyond this simple single reflection-axis setup to account for higher orders of reflection as well as rotational properties (so, ignoring colour change, a simple yin-yang motif can be described as exhibiting two-fold rotational symmetry, with one of the constituent components imagined to rotate through 180° to coincide with its partner). So, in principle, therefore, a motif, if imagined inscribed within a circle, can express much higher orders of reflection symmetry depending on the number of reflection axes passing through its centre and much higher orders of rotational symmetry also, depending on the angle of rotation of the smallest part. In the scientific and mathematical literature, a commonly used notation relies on the use of the letter *d* (for 'dihedral'); so motifs could be classified as d1, where one reflection action is present, d2 where two reflection actions are present, etc. Meanwhile, if a motif exhibits rotational characteristics, the commonly used notation relies on the use of the letter *c* (for 'cyclic'); so where rotation through 180° is present, the motif can be classified as c2; with three-fold rotation through 120°, the motif is classified as c3, etc.

It is apparent therefore that, where motifs are symmetrical, they may be classified by reference to their symmetry characteristics either in terms of their reflection or their rotation properties. Free-standing motifs, with no apparent symmetry characteristics, are a common design category found, for example, in embroidered textiles. Occasionally they form repeating patterns (often in repeating border form and, only occasionally, in the context of embroideries, as repeating all-over patterns). By their very nature, all regular patterns exhibit repetition, whereby a motif or other repeating unit is reproduced (or 'translated') on a regular basis across the plane in one or more directions, to produce regular border patterns or regular all-over patterns respectively.

5.4   SYMMETRY IN PATTERNS

As noted previously in this chapter, regular border patterns display translational symmetry in one direction across the plane, between two imaginary (or, occasionally, real) parallel lines. From the seven possible symmetry combinations, one involves translation only and the other six combinations of two or more symmetry operations. Meanwhile, regular all-over patterns display translational symmetry in two directions across the plane, and seventeen possible symmetry combinations unfold, one consisting of translation only and the others with combinations of two or more of the four symmetry operations. Both regular border patterns and regular all-over patterns can be classified with respect to their symmetry combinations. Schematic representations of symmetry combinations in motifs, and regular border and all-over patterns (not given here) together with an explanation of the most commonly used notation can be found in Hann (2012, pp. 72–96).

5.5   CLASSIFICATION, COMPARISON AND
THEORETICAL DEVELOPMENT

As noted previously in the introduction to this chapter, classification is a fundamental activity when explaining design types such as patterns. It offers avenues to identify and understand, yet it is a difficult challenge. Where a system of classification (or cataloguing) does not allow reproducibility from researcher to researcher, comparisons between designs are hampered. The development of symmetry classification was of importance for it offered a systematic and reproducible measure of design structure.

By the late-nineteenth century, there was a tendency for scholars interested in motifs and patterns to produce a Linnaean-type classification system (used often in a museum environment). This is indeed a crucial initial step, and an accurate taxonomic ordering and classification, based, for example, at the most basic level, on whether a motif or pattern is floral, animal, abstract, geometric, etc., can contribute to our understanding of the visual arts. In the textiles context, such classifications and descriptions have in many instances been extended to account for further characteristics such as the symbolism of motifs and patterns and the relevant processing technology. A further extension is however possible: consideration of the underlying structural characteristics (or symmetries) of the constituent motifs and/or patterns. Consideration of symmetry in patterns produced in cultures other than one's own can provide access to a dimension of meaning not readily obvious to cultural outsiders (Washburn 2004).

As noted above (Section 5.1), a key finding from the numerous research contributions is that, when a representative sample of designs from a given cultural setting (or historical period) is analysed with respect to its underlying symmetry characteristics and classified into the various symmetry classes (motifs with various orders of reflection or rotation, seven regular border-pattern classes and seventeen regular all-over-pattern classes), a non-random and unique distribution results, indicating that particular symmetry classes have been deliberately selected, and symmetry classification is culturally sensitive. Subject to the availability of representative data, symmetry analysis may help to pinpoint periods of cultural continuity and change. Further to this, symmetry analysis permits comparisons between data from various sources. For example, similar symmetry distributions shown by Sindhi embroideries (from Pakistan) and embroideries produced in east Africa may indicate the possibility of direct or indirect cultural transfer.

Washburn and Crowe (1988) is probably the best introduction to the subject. It succeeds in imparting a sophisticated understanding of symmetry concepts and terminology among the less mathematically gifted. Ways of classifying one and two-dimensional patterns, and their counter-change possibilities were explained by Washburn and Crowe (1988) and illustrated by reference to a large collection of historical patterns. Overall, a sophisticated anthropological approach was taken.

5.6 SUMMARY

This chapter explained the nature of symmetry and recognised that the principle of symmetry was of universal relevance across the visual arts, design, architecture, engineering and the physical and biological sciences. Important literature concerned with the development of knowledge relating to the subject was identified. It was seen that the symmetry characteristics of motifs, regular border patterns and regular all-over patterns could be considered. Four fundamental symmetry operations were identified: reflection, rotation, translation and glide-reflection, and it was shown how combinations of these symmetry operations could yield infinite varieties of motifs displaying reflection or rotation, seven regular border pattern varieties and seventeen regular all-over pattern varieties. Symmetry gives structure to designs on the one hand and acts as a constraining force on the other (so both regular border patterns and regular all-over pattern varieties are restricted geometrically in the number of possible underpinning structures associated with each). An important research result has been that, when representative collections of data for different cultural or historical contexts were analysed with respect to their symmetry characteristics, each set of data was found to express its own unique symmetry characteristics, indicating that symmetry characteristics are in some way culturally sensitive; because of their associated objectivity, results have highlighted the possibility that symmetry analysis is a technique which is reproducible from researcher to researcher. Commentary related to the symmetry characteristics of various images is provided as captions to Figures 5.2 to 5.11.

EXERCISES

5a   Classification of Symmetry in Finite Figures

Collect photographs or drawings of twenty automobile hubcaps (each from a vehicle of a different make/model). Classify each image collected with respect to its component symmetry characteristics using the notation *cn* or *dn* (where *c* represents a motif with rotational characteristics only, *d* a motif with reflection characteristics and *n* is either the number of rotations within 360° or the number of reflection axes).

5b   Identification of Symmetry Characteristics in Regular Border Patterns

Referring to the collection of regular border patterns created in response to Exercise 3b, identify and list the symmetry characteristics of each.

5c   Identification of Symmetry Characteristics in Regular All-Over Patterns

Referring to all regular patterns created in response to exercises given in Chapters 1 to 4 of this book, classify each with respect to its symmetry characteristics.

REFERENCES

Broug, E. 2008. *Islamic Geometric Patterns*. London: Thames and Hudson.

Broug, E. 2013. *Islamic Geometric Design*. London: Thames and Hudson.

Broug, E. 2016. *Islamic Design Workbook*. London: Thames and Hudson.

Conway, J. H., H. Burgiel and C. Goodman-Strauss. 2008. *The Symmetries of Things*. Wellesley (Mass.): A. K. Peters.

Endress, P. K. 1999. Symmetry in flowers: Diversity and evolution. *International Journal of Plant Sciences*, 160 (S6): S3–S23.

Farris, F. A. 2015. *Creating Symmetry*. New Jersey: Princeton University Press.

Fathauer, R. 2008. *Designing and Drawing Tessellations*. Phoenix (Arizona): Tessellations.

Grünbaum, B. and G. C. Shephard. 1992. Interlace patterns in Islamic and Moorish art. *Leonardo*, 25 (3/4): 331–339.

Hahn, W. 1998. *Symmetry as a Developmental Principle in Nature and Art*. Singapore, New Jersey, London and Hong Kong: World Scientific.

Hankin, E. H. 1934. Some difficult saracenic designs II. A pattern containing seven-rayed stars. *The Mathematical Gazette*, 18 (229): 165–168.

Hann, M. 2012. *Structure and Form in Design*. London and New York: Berg.

Hann, M. A. and B. Thomas. 2010. Recognition, differentiation and classification of regular repeating patterns in textiles. In *Modelling and Predicting Textile Behaviour*, X. Chen (ed.), 422–456. Cambridge: Woodhead Publishing.

Hann, M. A. and G. M. Thomson. 1992. The geometry of regular repeating patterns. *Textile Progress Series*, 22 (1): 1.

Hargittai, I. (ed). 1986. *Symmetry: Unifying Human Understanding*. New York: Pergamon.

Hargittai, I. (ed). 1989. *Symmetry 2: Unifying Human Understanding*. New York: Pergamon.

Jackson, P. 2018. *How to Make Repeat Patterns*. London: Laurence King Publishers.

Knight, T. W. 1998. Infinite patterns and their symmetries. *Leonardo*, 31 (4): 305–312.

Locher, J. L. (ed.) 1982. *Tessellations. M. C. Escher, His Life and Complete Graphics Works*. New York: Abradale Press.

McDowell, R. B. 1994. *Symmetry. A Design System for Quiltmakers*. Lafayette (CA): C & T Publishing.

Schattschneider, D. 1978. The plane groups. Their recognition and notation. *The American Mathematical Monthly*, 85 (6): 439–450.

Schattschneider, D. 1990. *Visions of Symmetry*. London: Thames and Hudson.

Schauermann, F. L. 1892. *Theory and Analysis of Ornament*. London: Sampson Low, Marston and Company.

Stevens, P. S. 1981. *Handbook of Regular Patterns*. Cambridge (Mass): MIT Press.

Swenson, C. 1978. The symmetry potentials of the ornamental pages of the Lindisfarne Gospels. *Gesta*, 17 (1): 9–17.

Walker, T. and R. Padwick. 1977. *Pattern: Its Structure and Geometry*. Sunderland (UK): Ceolfrith Press.

Washburn, D. K. (ed.) 1983. *Structure and Cognition in Art*. Cambridge (UK): Cambridge University Press.

Washburn, D. K. (ed.) 2004. *Embedded Symmetries. Natural and Cultural*. Albuquerque: University of New Mexico Press.

Washburn, D. K. and D. W. Crowe (eds) 2004. *Symmetry Comes of Age: The Role of Pattern in Culture*. Seattle and London: University of Washington Press.

Washburn, D. K. and D. W. Crowe. 1988. *Symmetries of Culture: Theory and Practice of Plane Pattern Analysis*. Seattle: University of Washington Press.

Woods, H. J. 1935a. The geometrical basis of pattern design. Part 1: Point and line symmetry in simple figures and borders. *Journal of the Textile Institute. Transactions*, 26: T197–T210.

Woods, H. J. 1935b. The geometrical basis of pattern design. Part 2: Nets and sateens. *Journal of the Textile Institute. Transactions*, 26: T293–T308.

Woods, H. J. 1935c. The geometrical basis of pattern design. Part 3: Geometrical symmetry in plane patterns. *Journal of the Textile Institute. Transactions*, 26: T341–T357.

Woods, H. J. 1936. The geometrical basis of pattern design. Part 4: Counterchange symmetry in plane patterns. *Journal of the Textile Institute. Transactions*, 27: T305–T320.

Yang, C. N. 1996. Symmetry and physics. *Proceedings of the American Philosophical Society*, 140 (3): 267–288.

Figure 5.1
(Left to right): Schematic illustrations of the four symmetry operations of translation, rotation, reflection and glide-reflection (drawing by Chaoran Wang).

Figure 5.2
The circular central portion of this plate shows reflection symmetry, whereas the outer portion shows a regular border pattern presented in circular format. Item held at the Metropolitan Museum of Art, New York (drawing by Haohong Zhuang).

Figure 5.3
Four-fold rotational symmetry forms the basis of the embellishments on this historical plate. Item held at the Metropolitan Museum of Art, New York (drawing by Haohong Zhuang).

Figure 5.4
Broadly, this figure (from an historical textile sample) shows reflectional symmetry at midway points both vertically and horizontally. Item held at the Metropolitan Museum of Art, New York (drawing by Haohong Zhuang).

Figure 5.5
This image, from an historical textile, shows (approximate) vertical reflection symmetry. Item held at the Metropolitan Museum of Art, New York (drawing by Haohong Zhuang).

Figure 5.6
This image, from an historical textile, appears to show (approximate) two-fold rotational symmetry but this is not the case. Item held at the Metropolitan Museum of Art, New York (drawing by Haohong Zhuang).

Figure 5.7
This image from an historical textile, suggests various symmetry characteristics though no formal repetition can be detected. Item held at the Metropolitan Museum of Art, New York (drawing by Haohong Zhuang).

Figure 5.8
This image, from an historical Turkish textile, shows approximate vertical reflection symmetry.
Item held at the Metropolitan Museum of Art, New York (drawing by Haohong Zhuang).

Figure 5.9
This image, from an historical textile, shows vertical reflection symmetry. Item held at the Metropolitan Museum of Art, New York (drawing by Haohong Zhuang).

Figure 5.10
An all-over pattern based on the experiments of Kepler. (Adapted from Nicolai, C. 2008. Grid Index. Berlin: Gestalten.)

Figure 5.11
An historical textile showing approximate two-fold rotational symmetry (drawing by Haohong Zhuang).

CHAPTER 6

GRIDS

6.1 INTRODUCTION

This chapter introduces various forms of space division, including formats suggested by common tiling arrangements, brick bonds, traditional Chinese lattice designs and stained glass window formats, all of value as compositional grids to visual artists and designers. It is believed that these have undergone much development over the years and are well resolved visually; importantly, when considered of no value in the past, they were probably simply removed from use.

6.2 SINGLE-PAGE GRIDS

The most common form of single-page grid is associated with the so-called 'rule (or law) of thirds' (Elam 2004, p. 13) consisting of a square or rectangle divided into thirds by two sets of parallel lines (one set intersecting the other at 90°); four intersections are created and these can act as points of 'optimal focus' in compositions created using this format (Figure 6.1). Elam (2004) explored this three-by-three structure and presented numerous examples of alternative formats taken from both student work and well-known twentieth-century posters. Comprehensive analysis and discussion were a feature, and Elam's work is to be strongly recommended to readers who may wish to explore the numerous alternatives offered to the designer when faced with the challenge of a square or rectangular composition and the necessity to include varieties of information in an organised manner.

Elam (2004, p. 13) commented: 'An awareness of the law of thirds enables the designer to focus attention where it will most naturally occur and to control the compositional space. Elements do not need to land directly on the intersecting point as close proximity draws attention to them'. Interestingly, the three-by-three framework can be turned to an angle of (say) 30°, to allow all components to be placed at a 30° orientation within the square format. Elam (2004) provided a useful and worthwhile text which complements the numerous examples of successful composition.

Ambrose and Harris (2008, pp. 50–51) considered the same three-by-three format and advised that 'it was a guide to image composition and layout' and that it could 'help to produce dynamic results', were lines intersected. They added that locating 'key visual elements in the active hotspots [where grid lines crossed] of a composition helps to draw attention to them, giving an offset balance to the overall composition. Positioning elements using the rule of thirds introduces proportional spacing into the design,

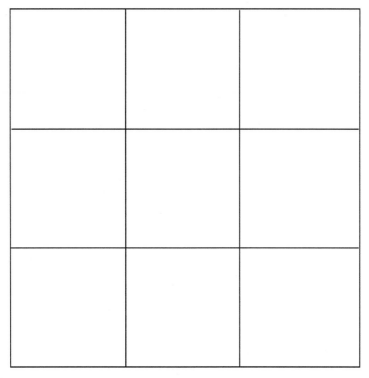

Figure 6.1
A single-page grid of the type associated with the 'rule (or law) of thirds' (drawing by Chaoran Wang).

which helps to establish an aesthetically pleasing balance' (Ambrose and Harris 2008, pp. 50–51). Interestingly, Puhalla (2011, p. 122) observed, when referring to 'the rule of thirds', that this was not a 'rule', but rather a 'rule of thumb' associated with capturing photographic images.

It should be noted that although Elam (2004) and Ambrose and Harris (2008) were concerned largely with single-page composition, it is suggested nevertheless that such perspectives should indeed be brought to the attention of all pattern-design practitioners, students and experienced professionals, as it is believed that awareness of these will help to inform compositional decisions relating to regular pattern design also.

Ambrose and Harris (2008) provided a comprehensive explanation for the use of one-page-type grids. They observed that such a grid was 'the foundation' upon which a design was constructed, and it allowed 'the designer to effectively organise various elements on a page' (Ambrose and Harris 2008, p. 6). Grids provide skeletal frameworks and 'bring order and structure to designs, whether they are simple...or as heavily populated as those on newspaper websites' (Ambrose and Harris, 2008, p. 6).

Ambrose and Harris (2008, p. 11) provided numerous illustrations, many from highly prestigious projects, and highlighted the fact that grids allowed designers 'to make informed decisions' and acted as frameworks to organise the components, providing 'accuracy and consistency in the placement of page elements'. Interestingly they observed that 'the structural principles behind designing of a printed page still apply [to an electronic format associated with the digital age], since the way we read a page [paper or screen] and how we extract information from it remains the same' (Ambrose and Harris 2008, p. 12). They commented further that the basic function of a grid was to organise information on a page and noted that the way this was achieved had been developed over centuries from 'simple pages of text, to the incorporation of images and to the diverse possibilities provided by modern design software' (Ambrose and Harris 2008, p. 12). In the organisation of information, they stressed the importance of 'hierarchy', which they considered to be a logical organisation of text and images with the position of each component dependent on its level of importance (Ambrose and Harris 2008, p. 12). Elam (2004) showed the use of grid systems in design organisation, paying attention particularly to the extensive organisational variation possible within the simple three-by-three grid.

6.3   REGULAR ALL-OVER GRIDS

The present author argued previously that various forms of tiling (or tessellation) were of potential value as compositional aids to visual artists and designers (Hann 2015, pp. 108–114). It has been noted that the three-grid forms (or tilings), referred to as 'regular', are created from associations of equal-size equilateral triangles or squares or regular hexagons (Figures 6.2 to 6.4); these were introduced previously in Chapter 4 of the present book. While these are the only three regular polygons capable on their own of tiling the two-dimensional plane without gap or overlap, it has been noted that various other possibilities are apparent through using two or more regular polygons as unit cells. Such arrangements are explained briefly below.

Figure 6.2
A regular grid consisting of equal-size equilateral triangles (drawing by Chaoran Wang).

Figure 6.3
A regular grid consisting of equal-size squares (drawing by Chaoran Wang).

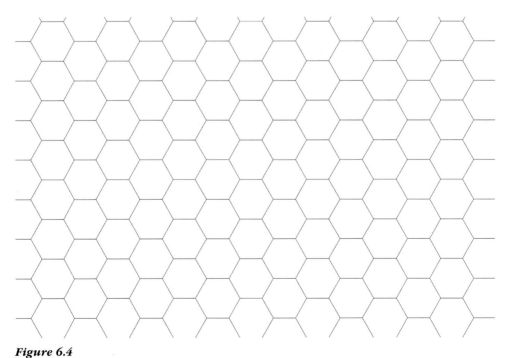

Figure 6.4
A regular grid consisting of equal-size regular hexagons (drawing by Chaoran Wang).

6.4 SEMI- AND DEMI-REGULAR GRIDS

A semi-regular grid (or tiling) consists of more than one type of unit cell. With each of these arrangements, the vertices (the points where the angles of each unit cell meet) are identical. A total of eight possibilities arise (Figures 6.5 to 6.12). A further form involving

(Text continued after figures)

Figure 6.5–Figure 6.12
Semi-regular grids.

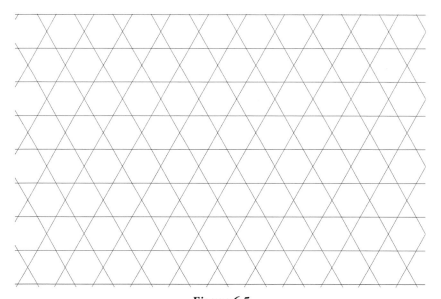

Figure 6.5

Figure 6.6

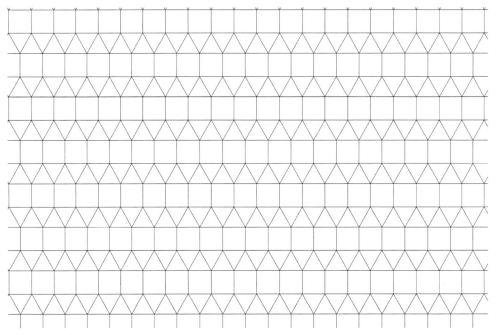

Figure 6.7

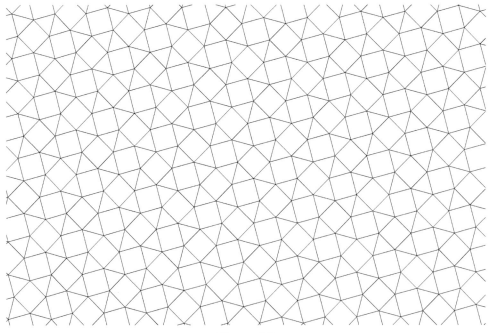

Figure 6.8

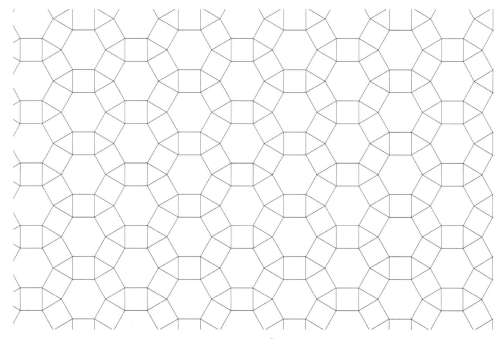

Figure 6.9

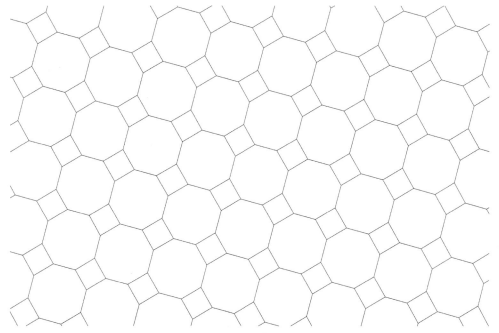

Figure 6.10

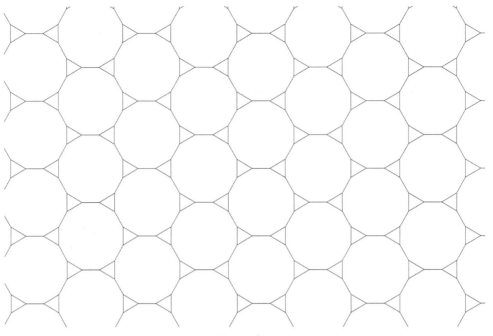

Figure 6.11

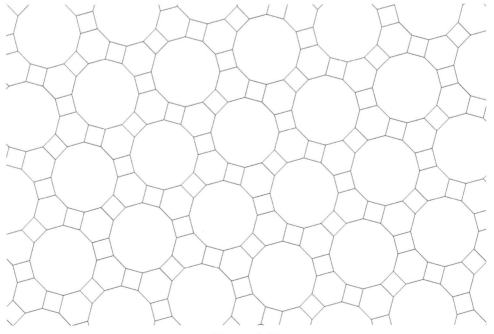

Figure 6.12

differing vertices is known as the demi-regular tilings and these too can be of value in the construction of regular patterns; examples were given previously by Hann (2015, pp. 110–111).

6.5 BRICK BONDS, LATTICES AND STAINED GLASS

Bricks can be used to create walls or to pave horizontal walking surfaces. With the former, structural characteristics are crucial as the construction may need to be load bearing, whereas, with the latter, aesthetic considerations may guide the arrangement of parts. To create an efficient, potentially load-bearing, construction, it is crucial to ensure that the joins in each row (or 'course') of bricks do not coincide vertically with the joints between brick courses above or below. Various arrangement of parts, often ensuring structural strength, have evolved over the years. The reason for introducing load-bearing arrangements here is to propose that they are potentially suited to pattern composition. Brick arrangements (or 'bonds', as they are called), in most cases, have evolved over centuries and unsuited arrangements were probably simply removed from use; the same was true, probably, also of arrangements used in pavement construction.

In the book *Patterns: Design and Composition*, Hann and Moxon (2019) argued that brick bonds offered guidelines in the construction of pattern design. Invariably, a combined rectangle and square grid is available, with squares invariably half the area of the rectangular face.

Stained-glass windows, found often in medieval European churches and other places of worship, consist of coloured pieces of flat glass, held together by strips of lead, arranged to suggest a composition that is often figurative in nature, invariably retelling in picture form some parable or event from biblical or similar sources. Similar types of stained-glass arrangements, often in domestic or other non-church settings, consist of clear glass components again separated by strips of lead. Often these were in square, rectangular or circular formats found to be pleasing aesthetically. It is proposed therefore that such arrangements (with stained or clear-glass panels) be reviewed and considered also as compositional grids to guide pattern composition.

Traditional Chinese lattice designs are probably among the most unexploited design source. These are the grid-type structures of wood, covered with mulberry paper on one side, and used as window or door frames in traditional Chinese houses. An extensive catalogue of types was produced in 1974 [1937] by Dye. Here are long-standing ready-made divisions of space. Although Dye's catalogue is focused only on lattices found or used in China, similar window and door coverings were used also in Korea and Japan. They provided some light from without but were not as effective in doing so as clear glass. Like tilings, brick bonds and stained-glass windows, lattices are a division of space, but use broader lines of division (around 2–3 centimetres of wood) than would be common in tilings (say 0.5 centimetres of plaster), brick bonds (say 1 centimetre of cement) or stained glass (say 1 centimetre of lead). Lattices divided the interior from the exterior in traditional households throughout much of China, Korea, Japan and a few locations elsewhere in Asia. Only shadows were visible from without or within. What was outside may have been imagined – maybe this gave rise to thoughts of dragons and other creatures in the night. Shadow-puppet performances may have a similar origin. Jali screens from India have similarities structurally, but were intended primarily to provide privacy in divided space. Catalogues for common structures used in Korea, Japan and India have not been found, but they may well exist in museums, libraries or family archives somewhere in Asia, Europe, North America or elsewhere.

Brian (1980) examined evidence for the historical distribution of brick bonds of various kinds in England up to 1800. These could be considered as successful compositions. Composition is the physical arrangement of components of a design or other visual statement. Success in composition indicates that the various components have been assembled in a way which is both visually (and practically) effective. All successful visual statements communicate a message and have a strong content and a focal point. Each visual statement and each composition is different however, so there are no set ingredients. The objective is always to hold the attention of the viewer, and there are different ways of achieving this. Scale and hierarchy are of importance. Different components of a design will be considered to have different levels of importance. So, in order to stress the importance of a particular component, it may be represented as larger and more brightly coloured ,while a component of less importance may be smaller and less boldly coloured. Never consider the space around components of a design to be of no value. Gaps between components of a design can bring attention to them and also allow them to breathe. Position of components is of importance. Never position these at random, expecting that they will appear to be favourably placed. Alignment is a safe technique which helps to avoid confusion. Alignment to the left can create neat and effective composition.

6.6   SUMMARY

From the viewpoint of the visual artist or designer, a grid is considered as a scaffolding, skeleton or framework on which to place components of a visual statement or design. Two-dimensional grids can either be considered as consisting of various shapes arranged edge-to-edge or as frameworks made up of unit cells, arranged on a (generally) flat plane, created by the intersection of series of parallel lines. In its simplest from, a grid is the division of a single polygon such as a square or rectangle, typically with two series of parallel lines overlapping at 90°. Where the lines cross, it may be that points of placement for components of a design are created or it may be that rectangles are created in which to place blocks of text or other components. Such grids are typically used in magazine or newspaper page design, indicating where to place text and images.

The rule of thirds is a technique where designers divide their designs into three rows and three columns, involving placing two parallel lines vertically and another two parallel lines at 90° horizontally. Where lines intersect, good focal points are created. As a rule, components that appear heavier in an image should be allowed to rest at the lower region of a composition. Also, probably most importantly, all images should engage the eye and mind of the viewer and keep attention within the image for as long as possible. Arrows or other directional devices, pointing to the perimeter of the image should be avoided, as these may lead the eyes of the viewer out of the image.

The components of a grid are the lines and the intersections. The lines create unit cells (each with at least three intersections and three lines). Each unit cell has three components: edges (or sides), a face (or two-dimensional shape), and vertices (where interior angles of adjacent unit cells meet). A grid comprises a series of intersecting lines. Usually two dimensional in nature, a grid is a framework to organise graphic elements of a design in relation to each other. The seminal work on the use of the grid was produced by Müller-Brockmann (2008), and this helped to spread its use initially in Europe and later North America. A grid is a scaffolding, skeleton or framework on which components of a design or other visual statement can be placed. There are numerous sorts of grid, though all conform to certain geometric rules. As mentioned numerous times in this present book, obvious selections for grid application, when it comes to regular patterns, are frameworks which conform to the three tiling arrangements known as the 'Platonic' or 'regular' tilings, composed of equal-size squares, or equal-size equilateral triangles or equal-size regular hexagons. Other tiling arrangements known as 'semi-regular' or 'demi-regular' tilings, both of which combine more than one regular polygon type, offer great potential also to pattern composition. A grid over two-dimensional space is the ideal framework to arrange a regular pattern, with each unit cell or group of unit cells holding a full repeating unit. Grids used for pattern composition provide a regular organisational framework, permitting all component parts to be in the most desirable place and ensuring that all regular patterns have a precise regularity of repetition demanded by modern forms of production.

Grids are of value to regular pattern designers and have been employed to ensure precision of repetition for centuries. In fact, it is envisaged that the design of all regular patterns on all surfaces requires careful consideration, planning and measurement, and grids offer frameworks which permit these activities. The majority of explanatory publications concerned with the use of grids in the visual arts and design are focused on the needs of 'graphic designers'. In this context, knowledge of grids is considered a necessary prerequisite to the creation of well-balanced non-repeating designs, including magazine, book, newspaper, promotional leaflet/brochure, poster and webpage composition.

So, this chapter explains further the nature of single-page grids (used in decisions relating to magazine, newspaper or webpage spreads) and the traditional regular all-over grids (of equilateral triangles, squares and regular hexagons) used in regular pattern organization and construction. The nature of various so-called 'semi-regular' possibilities are introduced, and the applicability of traditional brick bonds, Chinese lattices and stained-glass arrangements to possible pattern construction is highlighted. Figures 6.13 to 6.91 depict a wide variety of images from historical and other sources. Many are suited as frameworks to plan regular patterns.

EXERCISES

6a A Collection of Regular All-Over Patterns Using a Colour Palette of Your Choice

You are required to present a collection of six original regular all-over pattern designs, intended for a textile end use of your choice, using a means of production of your choice. Your collection should be based on a theme developed from a visit to a museum and should use a colour palette of no more than six colours. The sources used together with the anticipated means of production (e.g. printed, knitted, woven or felted), your selected colour palette and a statement of the fibrous raw materials to be used should be indicated in one A4 size mood/marketing board accompanying your collection. You are advised to keep the number of words to a minimum.

6b A Collection of Regular All-Over Patterns Based on Chinese Lattice Structures

You are required to present a collection of six original regular all-over patterns, intended for an end use of your choice. Your collection should be based on Chinese lattice structures presented in this book, and should use a palette of up to six colours. You are required also to present a single accompanying mood board (of A4 size) to indicate the visual sources used, how your collection was developed, the anticipated end use, the means of production and the raw materials and colour palette to be used.

6c A Collection of Regular All-Over Patterns Based on Stained-Glass Windows

You are required to present a collection of six original regular all-over patterns, intended for an end use of your choice. Your collection can be based on stained-glass-window structures from any source you may choose. You should use a palette of up to six colours. Also, present an accompanying single mood board (of A4 size) to indicate the visual sources used and the development of your collection as well as the anticipated end use, means of production and raw materials and colour palette to be used.

6d A Collection of Regular All-Over Patterns Based on Brick Bonds

You are required to present a collection of six original regular all-over patterns, intended for an end use of your choice. Your collection should be based on brick bonds from any source you may wish. You may use a palette of up to six colours. You are required also to present an accompanying single mood board (of A4 size) showing a sample of the visual sources used in the development of your collection, the means of production, raw materials and colour palette to be used and the anticipated end use.

6e A Collection of Regular All-Over Patterns Based on Semi-Regular Tilings

You are required to present a collection of six original regular all-over patterns, intended for an end use of your choice. Your collection should be based on a selection of semi-regular tiling formats (found in a source of your choice). Your colour palette should contain no more than six colours. You are required also to present a single accompanying mood/marketing board (of A4 size) showing a sample of the visual sources used, stages in the development of your collection, the means of production, raw materials and colour palette to be used and the anticipated end use.

REFERENCES

Ambrose, G. and P. Harris. 2008. *Basics. Design 07: Grids.* Lausanne: AVA Publishing.

Brian, A. 1980. The distribution of brick bonds in England up to 1800. *Vernacular Architecture,* 11 (1): 3–11.

Dye, D. S. 1974 [1937]. *Chinese Lattice Designs.* New York: Dover and, previously, Harvard (Mass.): Harvard University Press, under the title *A Grammar of Chinese Lattice* (2 vols.).

Elam, K. 2004. *Grid Systems: Principles of Organising Type.* New York: Princeton Architectural Press.

Hann, M. A. 2015. *Stripes, Grids and Checks.* London: Bloomsbury.

Hann, M. A. and I. S. Moxon. 2019. *Patterns: Design and Composition.* New York: Routledge.

Müller-Brockmann, J. J. 2008. *Grid Systems in Graphic Design: A Visual Communication Manual for Graphic Designers, Typographers and Three Dimensional Designers.* Sulgen/Zürich: Niggli.

Puhalla, D. M. 2011. *Design Elements: Form and Space.* Beverley, MA: Rockport.

Figure 6.13– Figure 6.28
Various tilings, brick bonds and lattice arrangements, all suited for development as compositional frameworks (drawings developed by Jia Zhuang).

Figure 6.13

Figure 6.14

Figure 6.15

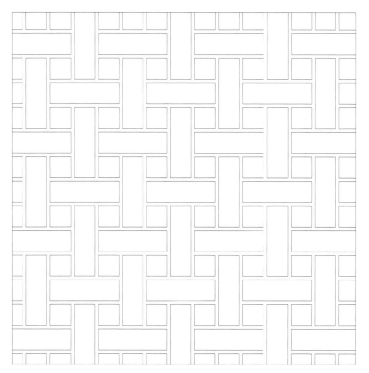

Figure 6.16

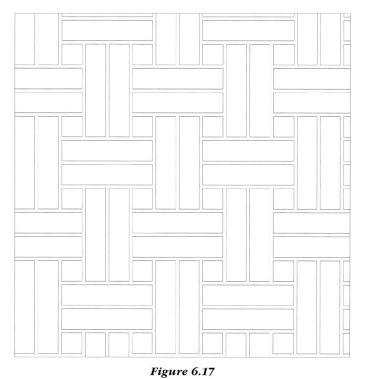

Figure 6.17

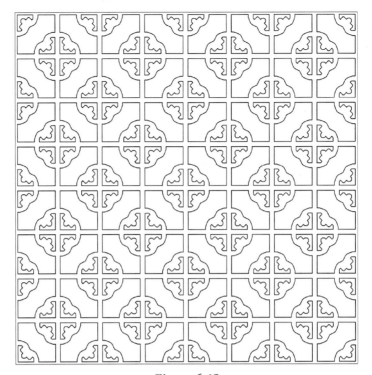

Figure 6.18

Figure 6.19

Figure 6.20

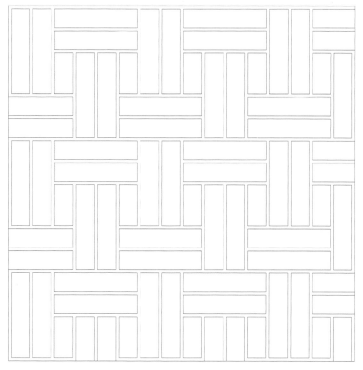

Figure 6.21

Figure 6.22

Figure 6.23

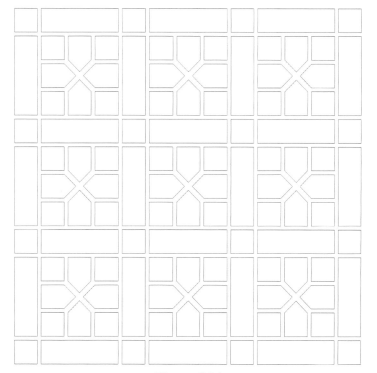

Figure 6.24

Figure 6.25

Figure 6.26

Figure 6.27

Figure 6.28

Figure 6.29–Figure 6.46
A collection of lattice arrangements, all adapted from Dye (1974 [1937]) and suited for further development as compositional frameworks (drawings by Haohong Zhuang).

Figure 6.29

Figure 6.30

Figure 6.31

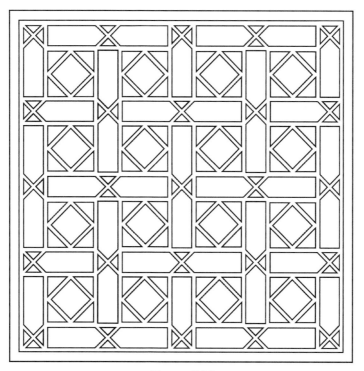

Figure 6.32

Figure 6.33

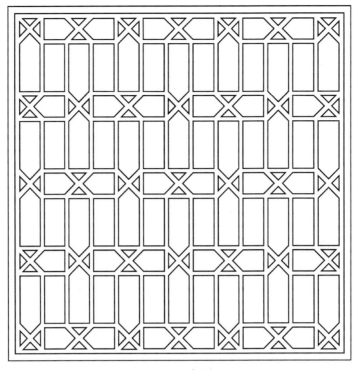

Figure 6.34

Figure 6.35

Figure 6.36

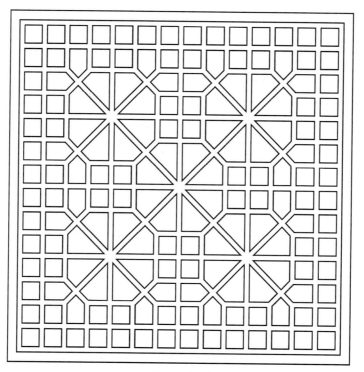

Figure 6.37

Figure 6.38

Figure 6.39

Figure 6.40

Figure 6.41

Figure 6.42

Figure 6.43

Figure 6.44

Figure 6.45

Figure 6.46

Figure 6.47–Figure 6.55

Adapted from Persian carved stucco designs, depicted in Dowlatshahi (1979), based on various grid types (drawings by Haohong Zhuang).

Figure 6.47

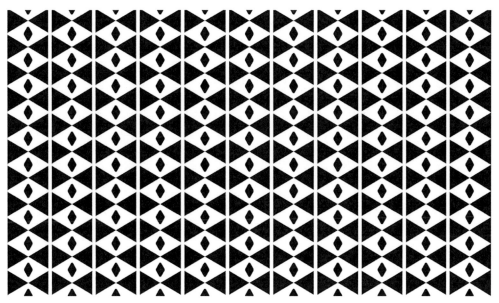

Figure 6.48

Figure 6.49

Figure 6.50

Figure 6.51

Figure 6.52

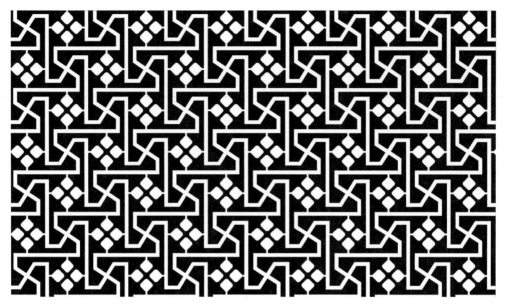

Figure 6.53

Figure 6.54

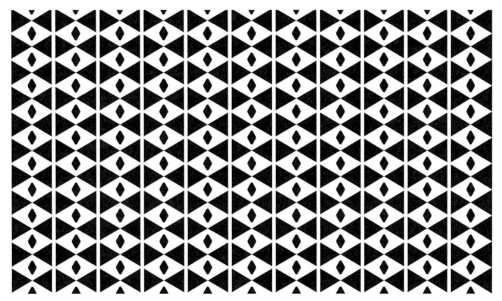

Figure 6.55

Figure 6.56–Figure 6.76
Images adapted from Gillon (1969), suited for further development focused on various end uses.

Figure 6.56

Figure 6.57

Figure 6.58

Figure 6.59

Figure 6.60

Figure 6.61

Figure 6.62

Figure 6.63

Figure 6.64

Figure 6.65

Figure 6.66

Figure 6.67

Figure 6.68

Figure 6.69

Figure 6.70

Figure 6.71

Figure 6.72

Figure 6.73

Figure 6.74

Figure 6.75

Figure 6.76

Figure 6.77– Figure 6.81
Various stained-glass and similar divisions (photographs taken in Belgium in 2016).

Figure 6.77

Figure 6.78

Figure 6.79

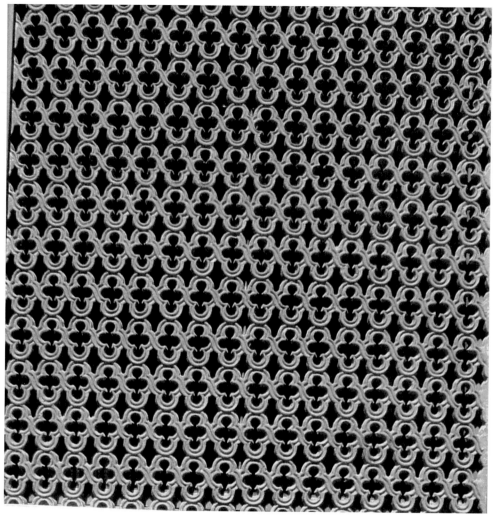

Figure 6.80

Figure 6.81

Figure 6.82–Figure 6.91

Regular all-over patterns produced by students at Asia University (Taiwan) in 2018. All designs based on grids. Contributions were made by Cindy Chao, Hsieh Hsin Ying, Li Yu Ling, Nancy Dai Yi Jie and Rachel H. Leung.

Figure 6.82

Figure 6.83

Figure 6.84

Figure 6.85

Figure 6.86

Figure 6.87

Figure 6.88

Figure 6.89

Figure 6.90

Figure 6.91

CHAPTER 7

CONCLUSION

7.1 INTRODUCTION

Regular all-over patterns consist of units of repetition, referred to often as motifs, which are reproduced by placing exact copies at precise distances vertically and horizontally so that the resultant all-over effect consists of a unit undergoing repetition covering a surface (invariably flat and referred to as a plane). Units of repetition are identical in size, shape and content and consist of exact combinations and distributions of lines, dots, textures and colours. Regular all-over patterns can be built on scaffoldings known as grids, each consisting of a regular distribution of squares, equilateral triangles, regular hexagons, or derivations or combinations of these or other geometrical figures. Such grids may be referred to as 'compositional grids', consisting of regularly repeating unit cells, with each cell acting as housing for a repeating component of the regular pattern. A compositional grid acts, therefore, as a guideline for the placement of component parts, and such grids may be apparent or hidden in the pattern's final depiction. Symmetry as a concept has attracted the attention of observers from a multitude of disciplines, including the visual arts and design, though with origins in various scientific areas of enquiry. All regular all-over patterns can be classified also with respect to their symmetry characteristics into one of seventeen symmetry groups based on one or more of four symmetry operations or geometrical actions. Such knowledge can inform the designer during the design process.

  It is the contention here that all the frameworks suggested (tilings, brick bonds, stained glass and lattice arrangements) are worth further consideration and, with minor adjustments, are suited as compositional grids for use by visual artists and designers. These frameworks have been developed, often, over many centuries and, although on the face of it their function is different from that which may be considered useful to twenty-first-century visual artists and designers, they nevertheless are worth attention as they show divisions of space which were technologically and functionally suited to their time of use. A series of propositions concerning regular pattern design and composition (principally for the attention of student readers) is presented below. Some may prove of value in portfolio preparation and others when completing and submitting a project.

7.2 MIND THE GAP

Sometimes two-dimensional designs, especially regular patterns, exhibit a foreground and background, with the components considered to be most essential placed in the foreground. It should be remembered, however, that the space around (i.e. the background) is just as important in compositional terms as the foreground. Where foreground and background exist, the line between the two should be clear, with no areas of fuzziness. In regular patterns with a clear foreground and background, the gap (or background) between motifs is a crucial part of the message; care should be given to determining the proportion of foreground to background (as changes in this proportion can alter the perception of the design radically). It should be recognised, however, that often, several colours will be present in the one regular pattern, so several layers of colours are occasionally of relevance. Especially here, care will need to be taken in the selection of colours for each individual component, and the pecking order of each component will need to be decided in advance.

7.3 SIDE BY SIDE

Successful regular pattern design is based on ensuring visual interaction between adjacent motifs, thus attracting the attention of the viewer for longer than otherwise. The focus should be on allowing the viewer's gaze to circulate around what is offered rather than allowing completion through a quick glimpse.

7.4 SYMPATHY

When designing a regular pattern, the process should be focused continually on ensuring sympathy with an anticipated end use, the method of manufacture, the colour palette, size of motifs and types of raw materials used; these should never be arbitrary selections at the end of the design process. The addition of a regular pattern should always enhance and not detract.

7.5 SIMPLICITY

The simplest interpretation is preferred, and additions to your work which raise unanswered questions should not be made. Do not annoy the viewer or reader by adding ambiguity. Order and clarity are preferred to disorder and fuzziness. Keep words and added text to an absolute minimum.

7.6 ON TIME AND NEVER LATE

Do not submit extra or less. Only submit what is required, no more and never less. Always hit the deadline, not before and never after. As you become more experienced, you realise that projects take longer to complete. It is not that you are slowing down but, rather, that you are becoming more critical of your work. While a student, allow an extra 10% of time towards the production of each successive collection of fully resolved images. Leave time to reconsider and to edit solutions.

7.7 DON'T TRY TO FOOL THE VIEWER

A straight line is only a straight line when it is a straight line. When crooked, do not try to fool yourself (or the viewer) that it is not. Always be precise. The viewer will become aware of flaws, even if only unconsciously. So, ensure that circles are circles, curves are smooth and lines are straight where intended.

7.8   PLAN

Good design is not achieved through random decisions, but instead is based on thorough planning, including market, colour and broader visual research. This brings to mind the maxim (often used by carpenters): measure three times, cut once.

7.9   COLOUR IT CORRECTLY

At any given time, different product markets and different geographic regions demand different colour palettes. Colour palettes at the forefront of demand, in each market, in each year, may slip from that position in the next year. A collection of designs coloured 'wrongly' (i.e. with a colour palette not in demand) will not sell.

7.10   ANTICIPATION

Design for the future. Never for the past or present. Good designers can anticipate what will be of importance culturally in one year's time and should design with this knowledge in mind. Product and market research should always accompany the preparation of a design collection. Be sure to take cognisance of all anticipated innovations and future legislation, especially relating to sustainability, raw material availability and environmental concerns, as well as anticipated economic forces.

7.11   THE KNOCK-ON EFFECT

Each successive stage of design development has a knock-on effect upon the next stage. The knock-on effect may be fortuitous or may present unanticipated problems. The cumulative effect, resultant from problems solved at each stage, allows the final design to emerge.

7.12   FUNDAMENTALS STILL REQUIRED IN THE DIGITAL AGE

During the second decade of the twenty-first century, there was a tendency among visual-arts-and-design students to believe that, with the onset of sophisticated digital technology and associated computer-enhanced design possibilities, there was no longer the need to become familiar with visual fundamentals. A contrary view is taken here, and it is stressed that without a well-developed knowledge of visual fundamentals, the basic ingredients of any visual composition (as outlined in Chapter 2 of this present book), the productive use and exploitation of modern-day computer graphics tools is not assured. In other words, it is emphasised that without a basic understanding of the fundamentals of visual composition, the fruitful exploitation of sophisticated digital-graphics software will not necessarily follow.

7.13   *LEARN TO KEEP THE RULES BEFORE YOU TRY TO BREAK THEM*

Often, for radical solutions to be arrived at, it is necessary to break the rules. However, before breaking the rules, it is necessary first to know what the rules are and how to keep them.

7.14   *MORE IS LESS*

Yes, the phrase attributed to Mies van der Rohe is 'less is more' (cited by Gombrich 1979, p. 17), but it is probably best to express this as 'more is less', a phrase particularly informative to students preparing a portfolio or response to an assignment. Weak work pulls down relatively strong work, so it is best to include only what is strong. Another way of expressing this is, 'a chain is only as strong as its weakest link'.

REFERENCE

Gombrich, E. H. 1979. *The Sense of Order.* Oxford: Phaidon.

GLOSSARY

This glossary is by no means fully comprehensive, but should prove of value to general readers, particularly those with little specialist knowledge of textile design, or those whose first language is not English.

Abstract: An abstract visual statement is where forms are unrecognisable and have no apparent visual reference to known entities. It is not based on a recognisable reality.

Acanthus: This is a leaf motif based on a plant of Mediterranean origin. The motif was used often in stylised form in architectural embellishment.

Accentuate: To ensure that an aspect of a visual composition is stressed or emphasised.

Achromatic: Often used to refer to a composition consisting of grey tones.

Acute angle: An angle less than 90°.

Adaptation: Where a past design has been modified and altered into a new format.

Aesthetic: This refers to the visual characteristics of a visual statement or design.

Airbrush: Imitates effects typical of those from a painter's spray gun.

Alhambra: The Alhambra Palace complex is in Granada, Spain, and is held to be one of the finest examples of Islamic architecture; it is renowned for its numerous tiling features.

All-over: A regular all-over pattern is where repeating units are evenly distributed across the plane.

Animal print: This is a regular pattern with a repeating unit resembling the fur or skin of an animal (e.g. tiger print or zebra print).

Arabesque: This is an elaborate design of curvilinear motifs, floral or geometric, often associated with Islamic visual arts or architecture.

Argyle check: A regular Argyle pattern is a regular all-over pattern consisting of diamond-shaped rectangles, often in knitted form in two or three colours.

Art Deco: This is an art style typical of the 1920s or 1930s, with a name derived from a Paris exhibition of 1925. Examples of the style, particularly in architectural form, can be found worldwide.

Art Nouveau: This is a design style associated with late-nineteenth century continental Europe, characterised by dynamic flowing curved forms seemingly derived from nature.

Arts and Crafts Movement: This is a design movement originating in nineteenth-century Britain. The stress was on craftsmanship with the rejection of industrial manufacture. The movement was associated with designers such as William Morris and C. F. A. Voysey.

Asymmetrical: These are forms without reflection or rotational symmetry.

Awning stripes: These are relatively broad stripes, usually vertical and of one solid colour against a lighter ground, with a design bearing a close resemblance to fabrics used for awnings.

Avant-garde: A term applied to an artist or a style of a finished work considered outside the conventional and radically different from the norm.

Background: Known also as 'ground', this is the part of a design which the viewer perceives to be farthest away and behind or between the main objects of interest in any composition.

Baclava: This is a diamond-shaped motif with saw-toothed or serrated sides used often in carpet design. The name is given also to a Turkish pastry cut in diamond shapes.

Basket weave: This is a regular all-over pattern, woven to imitate the interlaced construction of a straw basket.

Batik: The word is used to refer to both a product and the process. Batik is produced by a resist-dyeing technique, employing wax to cover areas of a cloth which, when immersed in a dyebath, will only take up colour in the regions not covered with wax. Some batiks show a peculiar veined effect caused by colour penetrating cracked wax areas of the cloth during the dyeing process.

Bauhaus: This is the name given to an early-twentieth-century German design movement.

Bead and reel: These often form a repeating component in architectural embellishments, combining a circular shape (a bead) and a cylindrical shape (a reel).

Bengal stripes: A woven or printed textile originating in India with stripes of similar width, alternating in light and dark colours. Known also as 'Regency' or 'tiger' stripes, the design was popular during the Regency period (second decade of the nineteenth century) in Britain and was used commonly in wallpaper, upholstery and shirting.

Bespoke: A one-off or one-of a kind product, made to order for a need. Custom-made for an event, person or situation.

Bird's eye: This is a small-scale, diamond-shaped weave, with a dot in the centre (deemed to suggest the eye of a bird).

Block printing: This is a type of relief printing, of ancient origin, employing carved wooden blocks.

Block repeat: This is a term used to refer to a composition where repeating units appear in horizontal blocks to the sides and directly above and below. Terms such as 'square repeat' and 'straight repeat' are used also to refer to the same arrangement.

Blotch: This is a term applied to a relatively large area of colour (often taking up a background position) in a design.

Border: A regular border (also known as 'frieze' or 'stripe') pattern is a pattern formed by the repetition of a repeating unit in one direction between two imaginary (or real) parallel lines.

Botanical: This is a term used to refer to a design with realistic representation of herbs, plants or other botanical motifs. Often such designs are based on botanical illustrations.

Boteh (also buta or Paisley motif): This is a pear-shaped motif with a small left or right twist at the narrow top. The motif was used typically in Kashmir textile manufacture and was adopted by Paisley (Scotland) textile weavers in the late-eighteenth century. The motif is believed to have initial origins in Persian floral design.

Bravais lattices: These are associated closely with pattern symmetry. Each of the seventeen all-over symmetry groups is associated with one of five varieties of lattice structures, each consisting of a regular series (or framework) of dots.

Brick layout: A composition in which each successive horizontal row of motifs is moved half of the repeating unit's width.

Buddhist symbols: There are a total of eight motifs under this heading, including the wheel, the lotus, the vase, the pair of fish, the canopy, the conch shell, the parasol and the endless knot (or knot of destiny). Various interpretations are given by different sources.

Camouflage: This is a design that conceals what it covers by 'blending in' to the surrounding environment.

Cartouche: A scroll-like device or medallion, often containing an inscription or an heraldic addition.

Celtic knot: Commonly associated with Celtic visual arts, knot designs appear as endless interlacings formed by a repeated path of one ribbon passing over and under another ribbon.

Check: A regular pattern, often woven, and based on overlapping squares.

Chevron: A pointed motif, often woven or printed, maybe forming part of a continuous zig-zag.

Chinoiserie: A European interpretation of an Asian design, often Chinese in origin.

Circle and related parts or terms: These include an arc, diameter, chord, disc, rotation, radius, segment, sector, tangent, circumference, convex, concave and curve.

Collage: An assembly of different images, including photographs, scraps of cloth, paper or other items, glued onto a flat surface. Collage was used by various artists, including Pablo Picasso, in the early-twentieth century.

Colourway: A further version of a given design using a different selection of colours.

Commercial: Aimed specifically at the demand of a known market.

Composition: This is how individual elements or parts of a visual statement are organised.

Computer-aided design: Where computer software assists with the design process.

Coral: A regular all-over pattern suggestive of growth.

Contour: The outline of a form.

Counter-change: A term used to indicate a regular interchange of colour in parts of a design. For example, a regular two-colour counter-change pattern is one where the colour of a component of the regular pattern (such as a motif) exchanges colour with another area of the regular pattern (such as the background) on a regular basis.

Coverage: The term is used occasionally in print design to indicate the fraction or percentage of the surface area of a printed textile covered with dyestuff compared to the unprinted 'background' area. So, a coverage of 80% would indicate that 80%, or 4/5, of the fabric surface area has been covered with dyestuff.

Croquis: A sample of a design not in formal repeat (though giving the impression that it is) indicating the colour palette to be used in the final design.

Diamond: A regular diamond pattern is a design with the repeating unit composed within a diamond-shaped unit cell (often simply a square grid reoriented or turned by 45°).

Diaper: A regular diaper pattern is a small-scale pattern, often geometric in nature with repeating units touching and sharing a common side.

Digital: The term 'digital design' is used to refer to a design generated/developed using computer-based software (such as Photoshop or Illustrator).

Directional: A regular directional pattern is where the repeating units are oriented or pointed in a given direction. Occasionally, the term 'directional' can be used to indicate designs aligned closely to recent fashion trends.

District check: A regular district check pattern (such as tartan) is a checked textile with a particular sett (or arrangement of yarn types) aligned closely or associated with uniforms identifying a particular Scottish estate.

Dog's tooth (or hound's tooth): The repeating component of a regular dog's tooth pattern consists of a central square with a triangle projecting outward at each side of the square.

Dome: A dome is a feature associated with roof constructions in ancient architecture and is explained conventionally as either half a sphere or as an arch rotated around a vertical axis.

Dynamic rectangles: Dynamic rectangles and the related term 'dynamic symmetry' is associated with Jay Hambidge (1926) and the publication *The Elements of Dynamic Symmetry*. These constructions are based on a square and its diagonal used to create successive root-2, -3, -4 and -5 rectangles. It was argued by Hambidge that these constituted 'dynamic symmetry' and were the basis of successful proportion in ancient Greek and Egyptian visual arts and architecture.

Dynamic symmetry: See dynamic rectangles.

Egg and dart: This is a repeating unit consisting of an oval-shaped component placed beside an arrow-shaped device.

Ethnic design: A term used to refer to the designs associated with a particular culture (historically or in modern times), invariably non-European.

Fall-on: This is a term used in printing to refer to one transparent colour falling on another to produce a third colour.

Field: This is the main part of a design, often (at least in the case of carpets) surrounded by a border.

Figure ground reversal: Where figure ground (or positive negative) reversal occurs there is a simple exchange of colour within two parts of the design. Another term used commonly is 'counter-change'.

Fleur-de-lis: A stylised version of a three-petal lily.

Floral: A design depicting flowers or similar natural components (including leaves, pods and stalks).

Forecasting: The process of predicting what will be in demand in the future.

Foreground: The part of a design which appears to the viewer to be placed at the front of a composition.

Formalism: Strict conformity to a prescribed form of composition or style.

Fret: Often in border (or frieze) form, a regular fret or key pattern consists largely of lines that meet at right angles. The design probably reached its highest stage of development in ancient times as the 'Greek key pattern', found in floor mosaics and domestic ceramics.

Frieze: A regular frieze (or border) pattern which displays regular repetition of a repeating component between two imaginary (or real) parallel lines.

Geometric: A regular geometric pattern is one that uses one or more known geometric shapes as a repeating component. Often such designs are referred to simply as 'abstract'.

Gingham check: A regular gingham check pattern depicts solid colour squares created by overlapping warp- and weft-way stripes of the same width.

Greek key: A regular Greek key pattern, also known as a regular fret pattern, consists of lines meeting at right angles. Although found in numerous cultures, it was in the form of Greek key patterns that the highest stages of development were reached in ancient times. See fret.

Grid: A skeleton or structure. Typically, a grid consists of two sets of equal width and equal distance parallel lines, with one set overlapping the other at 90° (maybe a square 'grid', where each unit cell is of an equal-size square). Numerous other grid types are possible, using differently oriented sets of parallel lines. Grids provide structural frameworks to organise the elements of a composition including regular patterns.

Guilloche: A regular guilloche pattern has a repeating unit consisting of interlaced curved bands occasionally forming a series of circles.

Hairline stripe: A regular hairline stripe pattern is regarded as the thinnest possible regular stripe, with the stripe effect created probably by only one warp-ways thread.

Half-drop: Best visualised in the context of a grid with square or rectangular unit cells, a half-drop grid is where each alternate column of unit cells is pulled down by half the unit cell's length. A regular half-drop pattern is where adjacent repeating units are placed in the unit cells of such a grid.

Hand: This is a term used often to indicate the 'style' associated with a particular artist or designer.

Herati: This is a stylised (either curvilinear or consisting of straight lines) rosette, maybe enclosed in a diamond shape and often found used in carpet design.

Herringbone: A design showing a zig-zag or chevron appearance when considered horizontally (from left to right), though organised in longitudinal stripe format when produced.

Ikat: An ikat design is created by covering portions of warp and/or weft threads with a dye-resistant material (such as banana leaf or polypropylene cord). On emersion in a dye bath, the dye will be absorbed only in those areas not covered with the dye-resistant material. Ikat is regarded as a resist-dyeing method.

Intaglio: A printing method in which the design is incised, often in metal or stone.

Interlocking: A regular interlocking pattern is one showing an arrangement where repeating components are engaged and fit closely together. Another term is 'regular interlacement' pattern.

Islamic design: A general term used often to refer to designs found in buildings or on other objects associated with the Islamic religion.

Jali: A pierced marble or other stone screen, found in parts of India.

Lattice: This term is used to refer to a grid or network of lines or, occasionally, a collection of regular points.

Layout: In the context of regular patterns, the term layout is used to refer to the composition of repeating units and how these are repeated.

Liberty style: This term refers to the design style associated with Liberty and Company.

Line: A line is the path traced by a point in motion. Lines may be straight, curved, broken, dotted or zig-zagged, though generally considered as straight unless stated otherwise.

Madder: This is a brown-red dye obtained, traditionally, from the roots of the plant *Rubia tintorum*. The dyestuff is occasionally referred to as 'Turkey red'.

Medallion: This is invariably disc, circular or oval in shape, and may form a repeating component of a regular pattern.

Moiré: These effects consist of wavy lines (or visual interference) associated with wet silk.

Monochrome: An image produced using shades of one colour only. Often in black, greys and white.

Mosaic: Mosaic designs sometimes show regular repetition but are often non-repeating. These are created by laying small coloured glass or ceramic squares (with sides of say 3 centimetres) side by side.

Motif: A term used often to refer to the repeating component of a regular border or all-over pattern. One or more motifs undergo regular repetition to create the whole regular pattern. Often, motifs are referred to as the 'building blocks' of patterns.

Negative space: This is a term used often to refer to the space between motifs in a regular pattern.

Notan: An expression for the balance between light and dark in a composition, considered of great importance in traditional Japanese visual arts and design.

Ogee: This is an onion-shaped outline or motif.

Optical art: This is a term used to refer to artwork which includes various visual effects, including the illusions of movement or depth, a sense of vibration or moiré-type effects.

Paisley: A popular tear-drop-shaped motif often found in textile form, associated with the town of Paisley in Scotland, and developed during the nineteenth century from the buta (a motif with origins in Kashmir and, previously, Persian floral designs). See boteh.

Palette: A term used to refer to the selection of colours used in a design.

Palmette: This is a term used to refer to a stylised palm leaf with classical Greek origins.

Patina: The covering that forms with age and use, often through prolonged exposure to air.

Pattern: A design considered to have repeating components. In this present book, the term 'regular pattern' is used to refer to a pattern which exhibits the regular repetition of an equal-size repeating component.

Pencil stripes: A regular stripe pattern with components considered to be equal in width to the lines created by a pencil and with the distance between lines often wider than the lines themselves. The stripe components are not as fine as hairline stripes or pinstripes.

Pheasant's eye: A term used to refer to a weave consisting of diamond-shaped repeating units, with a dot in the centre of each. This is a weave which is similar to, but bigger than, the bird's eye weave.

Photomontage: A single image consisting of a series of photographic images (combined in a similar way to a collage).

Pinstripe: A regular stripe pattern, with repeating stripe components slightly wider than hairline stripes, but narrower than pencil stripes.

Plaid: A term used commonly to refer to a checked or tartan textile, with vertical and horizontal stripes crossing at 90°.

Point: This is the simplest entity associated with the visual arts and design and is held to have only location with no dimensions. When represented graphically, a tiny circular dot is often used to indicate that location.

Polychrome: A work executed in several colours.

Polygon: This is a geometric figure consisting of lines (or 'sides'). When referred to as 'regular', each side is of the same length and each internal angle the same. Equilateral triangles are three-sided regular polygons, squares are four-sided regular polygons, pentagons are five-sided polygons and hexagons are examples of six-sided polygons.

Polyhedron: This is a three-dimensional geometrical figure with each face a polygon. Examples (known as the Platonic solids) include a tetrahedron (a four-faced polyhedron), a cube (a six-faced polyhedron), an octahedron (an eight-faced polyhedron), the dodecahedron (a twelve-faced polyhedron) and an icosahedron (a twenty-faced polyhedron).

Polka dots: The term regular polka-dot pattern is used often to refer to a regular pattern of repeating circular forms in a single colour against a contrasting background.

Pop art: This term is associated with an art movement of the 1960s, featuring everyday objects.

Positive negative: Also referred to as a 'figure ground' relationship where reversal of colour between two parts of a design is a feature. The equivalent term 'counter-change' is often used in a textile context.

Powdered: A term used occasionally to refer to a regular pattern consisting of a scattering of dots or other tiny motifs.

Primitive art: This is a term used to refer to simple naïve art considered, in the past, to be less sophisticated than art emanating from other 'mainstream' sources mainly in Europe and North America.

Quadrilaterals: These are four-sided geometrical figures and include squares (with equal angles and equal-length sides), rectangles (right angles, with adjacent sides unequal in length), rhomboids (oblique-angled parallelograms, with adjacent sides unequal), diamonds (squares turned by 45°), lozenges (with opposite sides equal and with two obtuse and two acute angles), trapezoids (with opposite sides parallel) and isosceles trapezoids (with one pair of opposite sides parallel and the other two sides equal in length).

Regency stripes: These regular stripes originated in India and were known also as 'Bengal stripes'. Refer to Bengal stripes.

Repeat: A term used to refer to the repeating component of a regular design or the measured distance between such components (equal to the translation distance).

Roman stripes: This term is used often to refer to bold, broad, multicoloured vertical stripes.

Root rectangles: A square and its diagonal are the basis of root-2, -3, -4 and -5 rectangles.

Rose window: A circular window, often in stained glass found in the façades of Gothic churches.

Rosette: This is a term often used to refer to a motif which can be inscribed within a circle.

Sacred-cut square: This is a construction where a square is constructed within a square.

Sateen: A term used to refer to a smooth weft-faced fabric (where the weft threads are predominant on the face of the repeating unit), free from twill effects.

Satin: A term used to refer to a smooth, warp-faced fabric (where the warp threads are predominant on the face of the repeating unit), free from twill effects.

Scale: This can either relate to anticipated size (after production) or to a regular scale pattern, depicting overlapping semicircles or similar shapes.

Scroll: A regular scroll pattern has a repeating unit consisting of a ribbon-like motif reminiscent of a partly rolled scroll of paper.

Serpentine stripe: A regular serpentine stripe pattern consists of wavy stripe components reminiscent of snake movements.

Shape: A shape is the external appearance, outline or silhouette of a design or other object, whether in two or three dimensions.

Stripe: A regular stripe has a characteristic feature of parallel bars or lines oriented in a vertical, horizontal or, occasionally, a diagonal direction. These bars may be of different thicknesses and the full regular pattern can be in several colours, so long as the order of repetition of stripe widths and colours is identical across the design.

Surface pattern: A regular surface pattern is a regular pattern intended for use on a two-dimensional surface. This term includes regular patterns destined for textile use as well as wallpaper and floor-covering designs.

Sustainability: this refers to the upkeep of the environment into the long term.

Swatch: A small piece of cloth (say, 10 centimetres by 10 centimetres) used as a sample, showing the distribution of colours. Often, a full repeating component of the regular pattern is not included, but a section of it.

Symmetrical: A term used when referring to objects or images which display symmetrical characteristics.

Tactile: Relating to the sense of touch. A tactile texture is in three-dimensional space and can be sensed through touch.

Taoist symbols: Associated with the Chinese philosopher Laotse, there are eight of these, each representing a spirit or entity. They are generally listed as fan, sword, staff and gourd, basket of flowers, castanets, flute, lotus bud or flower, and tube and rods.

Tapestry: A woven textile with a predominance of weft (typically of wool yarn) covering warp threads (maybe of linen or cotton yarn). Often these depicted some historical or political event. Such textiles were commonly used as wall hangings or as coverings for upholstery.

Tartan: A regular tartan pattern is a check pattern, consisting of two assemblies of parallel lines, one superimposed on the other at 90°. Scottish clan tartans famously display a checked feature, using differently coloured yarns in woven-fabric form. Often the sequence of colours and numbers of yarns in weft and warp directions is equal (known as a 'balanced' tartan where this is the case).

Tessellation: A regular pattern where the component parts cover the plane without gap or overlap (this is often considered to be a characteristic feature of diaper patterns).

Tree of life: Tree of life motifs are found often on Oriental carpets and embroideries from various dates and sources, and on eighteenth and nineteenth-century Indian hand-painted cotton textiles (known as kalamkari).

Trellis: A regular trellis pattern consists of a series of parallel vertical components and parallel horizontal components intersecting at 90°.

Trigram: Associated often with the yin-yang symbol, this is an arrangement of three parallel lines (or line segments) which can be arranged in eight different formats, held to represent heaven, earth, wind, fire, water, clouds, mountain or thunder.

Turnover: A regular turnover pattern is one in which the repeating component has been flipped horizontally or vertically.

Twill: A regular twill pattern shows a characteristic diagonal orientation.

Two-colour counter-change: This is where the colours in a two-colour design are reversed on a regular basis.

Two-directional: A two-directional regular pattern will have some of its features oriented in one direction and the remaining features oriented in another direction.

Unbalanced: A regular pattern is unbalanced when it appears incomplete or unfinished, possibly with components missing or with incorrect emphasis on parts.

Wallpaper: A regular wallpaper pattern is simply a regular pattern intended for application on a two-dimensional interior surface. The use of the term extends beyond wallpaper to include wall coverings of various kinds.

Warp-ways: A term used to describe the direction of the series of vertical or warp threads in a woven fabric.

Weft-ways: A term used to describe the direction of the series of horizontal or weft threads in a woven fabric.

Windowpane check: A regular windowpane check is a widely spaced check pattern resembling the format of window panes.

Yin yang: This is a circular motif consisting of two equal-size curved components, one dark and one light. The motif is considered to be Chinese in origin, and the two components are considered to represent opposite forces or characteristics: positive/negative, sun/moon, hot/cold, day/night, etc.

REFERENCES USED TO ASSIST WITH ILLUSTRATIVE MATERIAL

Bennett, I. 1981. *Rugs and Carpets of the World*. London: Ferndale Editions.

Bidwell, J. K. 1986. A Babylonian Geometrical Algebra. *The College Mathematics Journal*, 17 (1): 22–31.

Dowlatshahi, A. 1979. *Persian Design and Motifs*. New York: Dover.

Dudley, C. J. 2010. *Canterbury Cathedral. Aspects of its Sacramental Geometry*. Bloomington (IN): Xlibris.

Dye, D. S. 1974 [1937]. *Chinese Lattice Designs*. New York: Dover and, previously, as *A Grammar of Chinese Lattice*, Boston: Harvard University Press.

Gillon, E. 1969. *Geometric Design and Ornament*. New York: Dover.

Hann, M. A. and C. Wang. 2016. Symmetry, ratio and proportion in Scottish clan tartans. *The Research Journal of Costume Culture*, 24 (6): 873–885.

Nicolai, C. 2008. *Grid Index*. Berlin: Gestalten.

Stone, P. F. 1997. *The Oriental Rug Lexicon*. London: Thames and Hudson.

Wiemer, W. and G. Wetzel. 1994. A report on data analysis of building geometry by computer. *Journal of the Society of Architectural Historians*, 53 (4): 448–460.

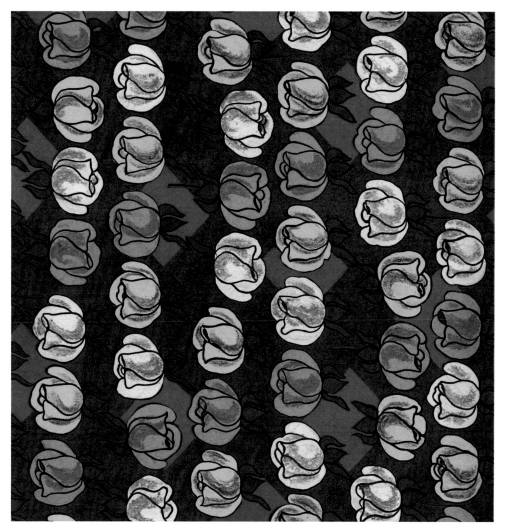

Plates 1–39

These plates depict designs in gouache on paper by the late Guido Marchini (1929–2009). Guido was born in Lombardy in northern Italy and, initially, trained in accountancy (an activity for which he had little desire). His development of early design skills was largely due to self-motivation, and was stimulated by working in the textile industry in Como. Ultimately, he went to Rome and worked for a leading textile designer of the day who encouraged him to develop his skills further by paying more attention to his independent work and, also, to travel to England 'to try his luck', particularly in mills around Manchester. Here, he acquired an agent thus ensuring that his work was produced by the best companies. After some years he settled in Stratford-upon-Avon and contributed with Peter Dingley to the development of a craft-oriented outlet, the first in the country devoted to British craft work. Guido was regarded as having 'an excellent eye' and this author rates him as one of the leading twentieth-century designers.

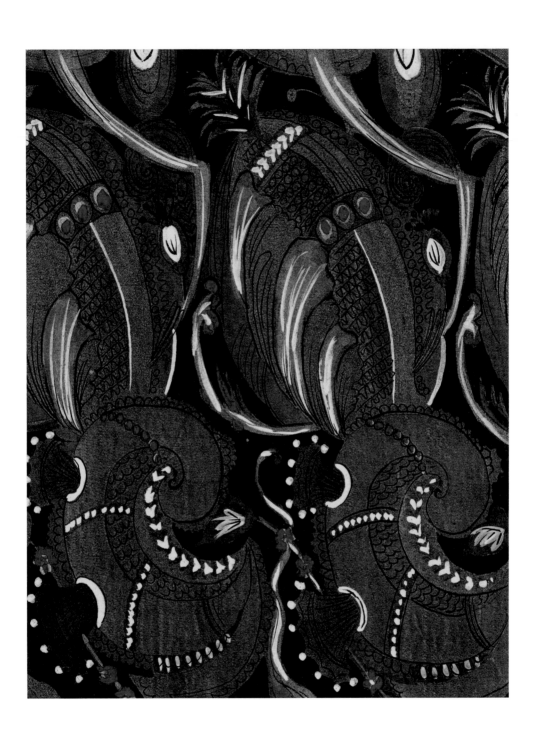

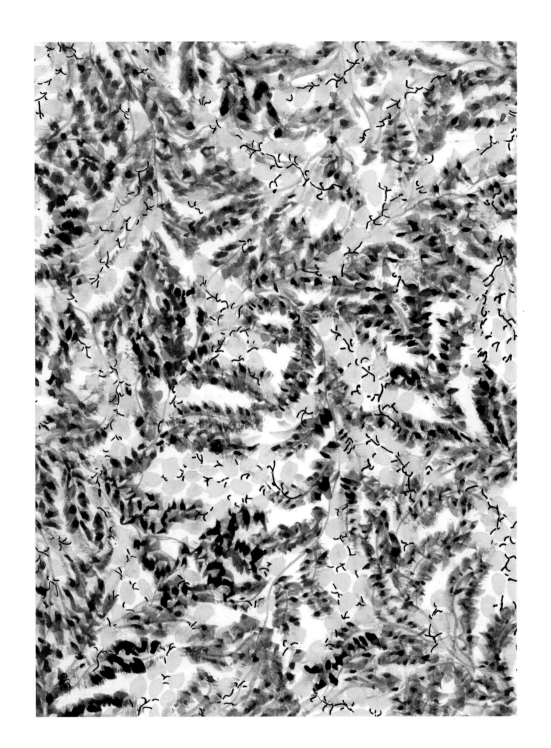

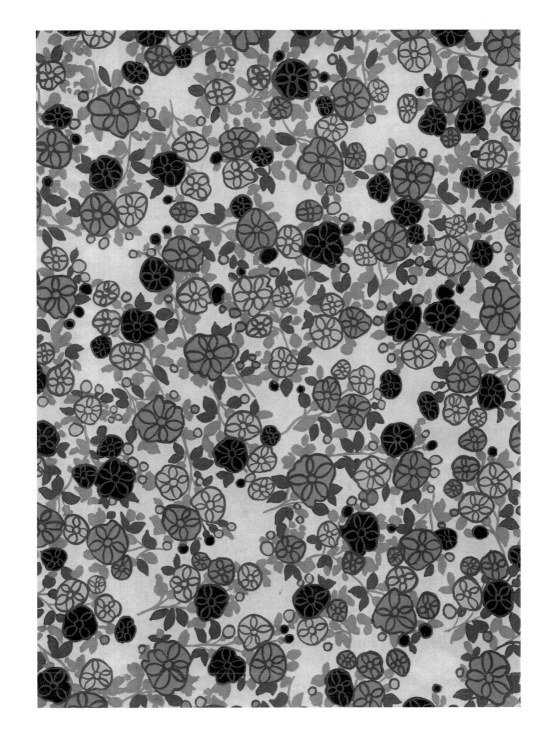

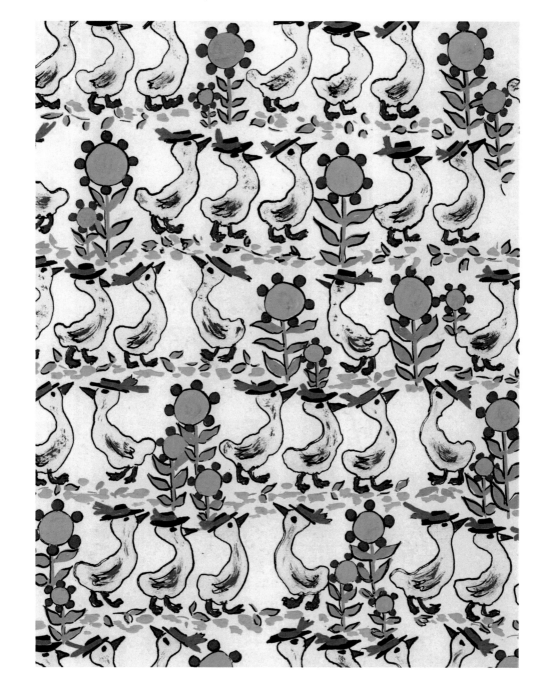

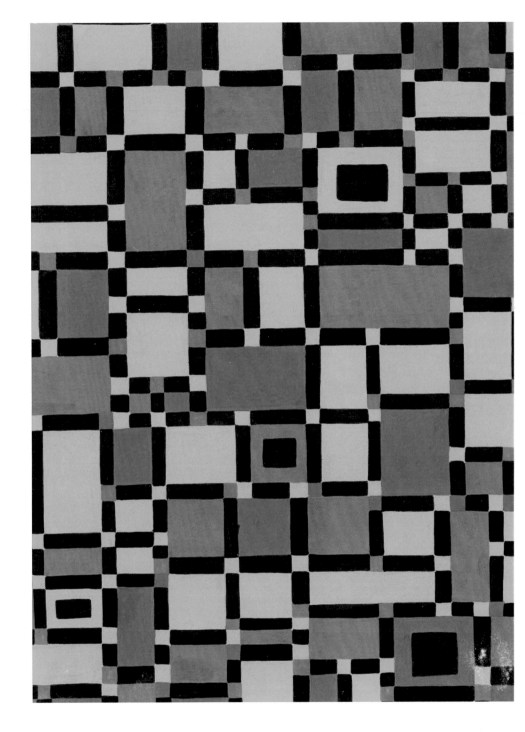

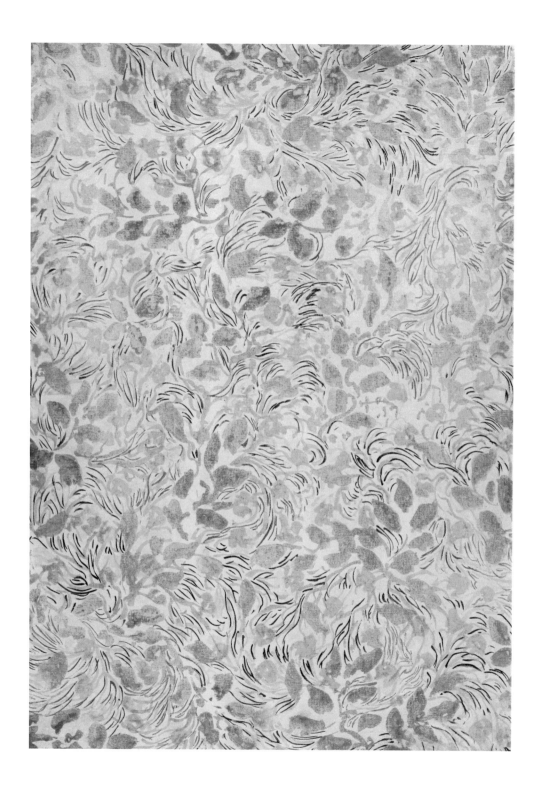

INDEX

Index

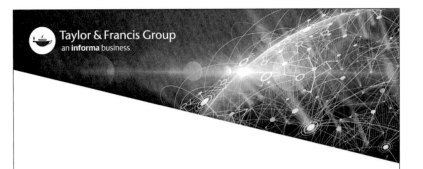

Printed and bound by CPI Group (UK) Ltd, Croydon, CR0 4YY

18/10/2024

01776236-0018